Routledge Revivals

Social Geography in International Perspective

Social geography has been one of the most important growth areas within the field of geography in recent decades. It has brought within geographical analysis a wide range of new topics, such as ethnic segregation, crime and environment and inner city problems. First published in 1986, this edited collection surveys the field of social geography. Using key international case studies from across Europe, North America, Sub-Saharan Africa and South Asia, authors discuss the different trends, leading figures and issues of concern within social geography throughout the world. This is a comprehensive and accessible study that will be of particular interest to students of social and human geography, urban and environmental planning.

Social Geography in International Perspective

Edited by
John Eyles

Routledge
Taylor & Francis Group

First published in 1986
by Croom Helm Ltd

This edition first published in 2014 by Routledge
2 Park Square, Milton Park, Abingdon, Oxon, OX14 4RN

Simultaneously published in the USA and Canada
by Routledge
711 Third Avenue, New York, NY 10017

Routledge is an imprint of the Taylor & Francis Group, an informa business

Publisher's Note
The publisher has gone to great lengths to ensure the quality of this reprint but points out that some imperfections in the original copies may be apparent.

Disclaimer
The publisher has made every effort to trace copyright holders and welcomes correspondence from those they have been unable to contact.

ISBN 13: 978-0-415-73457-8

Social Geography in International Perspective

Edited by
JOHN EYLES

CROOM HELM
London & Sydney

Croom Helm Ltd, Provident House, Burrell Row,
Beckenham, Kent BR3 1AT

Croom Helm Australia Pty Ltd, Suite 4, 6th Floor,
64-76 Kippax Street, Surry Hills, NSW 2010, Australia

British Library Cataloguing in Publication Data

Eyles, John
Social geography in international perspective.
1. Anthropo-geography
I. Title
304.2 GF41

ISBN 0-7099-0944-6

Printed and bound in Great Britain by Mackays of Chatham Ltd, Kent

CONTENTS

To Emrys Jones

Teacher, Colleague, Friend

PREFACE

Social geography as a term is now one hundred years' old. Its practice, however, covers mainly the last twenty-five years. During that time it has grown and developed rapidly and while there is perhaps still no overall agreement as to its nature, beyond the broad, but scarcely illuminating, 'social phenomena in space', it has been that part of the discipline at the forefront of reconstituting human geography as a social science and in the centre field with respect to the philosophical and methodological debates that have occurred over the past ten years.

It is, therefore, a fitting time to ask colleagues from different national geographies to reflect on and categorise the social geographies of their pasts, presents and futures. It is the time of a near coincidence of anniversaries: 1884 provided the first recorded use of the term in France and 1960 saw the publication of what may be regarded as the volume that put the subject on the map, Emrys Jones' A Social Geography of Belfast. Indeed to many of the contributors, Emrys has been teacher, colleague and friend for many years and on his retirement we collectively dedicate this volume to him as a fine social geographical scholar whose work will continue to inspire.

It is also the time when the Anglo-American and positivist dominances seem to be breaking down. Different, more place-specific ways of seeing and explaining are being considered and applied. In such a time, therefore, when different national geographies are examining more closely the specificities of local conditions, an international survey seems apposite. It is, therefore, a time to take stock and reflect on where we might go.

My final task is to thank those who have made this volume possible. First & foremost are the contributors themselves. For most, the task of setting down what must to them be commonplace is a matter of exegesis. I am glad they did, for I have already learnt much. I would also like to thank Carol Gray who, with great humour, produced excellent camera-ready copy. A special mention must also be made of Peter Sowden at Croom Helm, who, without panic or rancour, saw several 'final deadlines' pass by.

Introduction

DIFFUSION AND CONVERGENCE?

John Eyles

The purpose of this book is to explore the nature and development of social geography in selected countries and regions. Social geography is in fact one of the more recently developed subdisciplines within human geography. It appears, however, to be that part most closely connected with the social sciences in general and it has, therefore, become a significant dimension as human geography recasts itself as a social science. It remains a pertinent question though as to what degree this recasting is simply an Anglo-American phenomenon and one not pursued in the other 'geographies' of the world. Thus one of the main reasons for producing this volume was to try and answer this queston and thereby go some way towards correcting the academic myopia of much of the English-speaking world. Because of this desire no definitions or rigid guidelines were provided for the contributors. They were asked to describe the practice of social geography, the main influences, theoretical stances, key works and likely future directions as they saw them. With such a wide brief, it is not surprising that the contributors emphasised different things. It is indeed one of the most frustrating, yet also engaging and stimulating facets of editing a collection that contributors quite rightly refuse to write the papers the editor would write himself. Thus some emphasise the antecedents and present practices of their social geographies, while others are predominantly future-oriented. From all, however, we can discern the balance and relationships between theory, empirical work and policy orientation.

Contributors were asked to concentrate on the post-1945 period as it has been the last 40 years that has seen the major developments in social geography. The term 'social geography' is, however, now 100 years old, being first used by Reclus in France in 1884. This volume has, therefore, been somewhat fortuitously timed to coincide with that anniversary (almost!) and with the same degree of precision with the 1984 Paris conference of the International Geography Union, for which a summary statement on post-1945 world geography was produced (Johnston and Claval, 1984). It seemed to be useful to juxtapose social geographical developments with those of the whole subject. The contributions in this volume are, of course, far more narrowly based that those of the Johnston-Claval collection. The main reason for this is that social geography is not practised in every country or region of the world. Indeed even where it is found, its formalisation has not yet occurred. Thus in the United States, the Association of American Geographers has a socialist but no social geography study group. In Eastern Europe, the subdiscipline is subsumed under the more acceptable designation of 'economic geography'. At the other extreme, both the Scandinavian and Low Countries use the term to denote 'human geography'.

It has in fact not been an easy task to find contributors for those parts of the world that it was felt should be covered. The pressure of other commitments has meant that the volume itself has been a little delayed. Further, advice taken from various specialists working in and on different regions resulted in the exclusion of, for example, Latin America and Italy. It is interesting to see this exclusion justified for the former by Vila-Valenti (1984) who also points to the increasing importance of Anglo-American ideas and methods (spatial, scientific and marxian) in the Spanish-speaking world of geography. In Italy, social geographic themes are just beginning to emerge. Indeed, there appear to be three separate strands which owe much again to Anglo-American formulations but have parallels with current practice in Germany (see Thomale in this volume). A framework of marxist debate, if not analysis, has been established (see Quaini, 1974) while the welfare approach has been employed to describe contemporary levels of living in Italy (Zerbi, 1984a, b). The latter relies particularly on the application of quantitative methods in Italian geography as does, to some extent, the third strand that of perception studies (see Corna-Pellegrini, 1980).

Such omissions are not, however, serious. This book examines the social geography already being practised rather than that to come. Two more serious gaps must be mentioned, namely South-East Asia and the Netherlands. With respect to the former, the omission means an underemphasis of the processes and consequences of urbanisation. It may be argued that this work has mainly been carried out by non-native geographers employing Western models of development. While such endeavour has been criticised for the past ten years or so, research more sympathetic to local conditions and effects is only just beginning to emerge. In any event, Bradnock's chapter on South Asia covers that part of the region where the greatest number of indigenous, practising social geographers are to be found. The Netherlands has a much longer tradition of social geographical practice (Vooys, 1950). It has produced a number of influential scholars and papers. For example, Paasen (1976) examined the nature of human and social geography in terms of existential anthropology, a theme implicitly explored by the Chicago ethnographers who have been particularly influential in British social geography and one also pertinent to the methodological practices of humanist geographers, presently best represented in Britain and the United States. Further, Laan and Piersma (1982) contribute significantly to the debate on geographical enquiry by examining 'the image of man' (sic) in different theoretical contexts. But not only does Dutch social geography has this strong philosophical dimension, it also possesses an important practical or policy perspective. Many social geographers have become professors of planologie (or physical planning) and have a research interest in urban process and form in particular. The omission of a detailed consideration of the Netherlands in this volume is in part overcome by the recent volume of Tijdschrift voor Economische en Sociale Geografie, namely 74 part 5 (1983), given over to deliberations on the origins, nature and development of the influential Utrecht school. The French and German influences are traced (Heslinga, 1983) and one of its notable strengths, residential geography, is fully discussed (Ginkel, 1983). The latter discussion in particular notes not only the Anglo-American influences but also draws attention to the specific features of Dutch geography and the Dutch housing system that have shaped its distinctive developments. It may be that Utrecht-based residential geography represents a specifically Dutch adaptation to the French tradition of regional social geography (see Claval in this volume). This adaptation does appear to have been taken further at Amsterdam where geography and sociology remain close, sociography as a form of spatial sociology developed, and research interest in social and cultural phenomena is strong. It is only in very recent times that the impact of Anglo-American and Scandinavian geography has been felt. The conclusion of Paassen (1984, 227) is, however, instructive, suggesting significant convergence in international geography:

> Human geography might well be torn in two directions. One will reflect the 'hard' course of analytical research directed to studies of the spatial dimensions of societal systems and of large-scale organisations; the other will be the 'soft' course, of qualitative, situational research focused on the many and pluriform realities of man ... Geography should remain a battleground for different approaches. It should, however, pay more attention to institutional power and culture. Recent developments of an institutionally-directed economic and of political geography offer new prospects, as do a few experiments in qualitative research. Out of these, a new, more relevant policy-orientated geography might recover some of the lost ground and the Dutch tradition of 'social geography' gain a new impetus.

Definitions

It is perhaps against the spirit of this book to provide a catch-all definition of social geography. The contributors were not given one by the editor and while some discuss the issue of social geographic context, none provide one. Definitions are, however, illuminating because they are context- and time-dependent, informing about the concerns of particular scholars and eras. They are also significant in what they do not say and their silences can be as evocative as their clarion calls to particular kinds of practice. With due apologies to those omitted by oversight, the last twenty-five years or so has produced eight definitions of social geography, seven of which are provided by geographers working in the Anglo-American tradition. These are:

> the identification of different regions of

the earth's surface according to associations of social phenomena related to the total environment (Watson, 1957, 482)

the study of the patterns and processes (required) in understanding socially defined populations in a spatial setting (Pahl, 1965, 81)

the study of the areal (spatial) patterns and functional relations of social groups in the context of their social environment; the internal structure and external relations of the nodes of social activity, and the articulation of various channels of social communication (Buttimer, 1968, 144)

the analysis of the social patterns and processes arising from the distribution of, and access to, scarce resources and ... an examination of the societal causes of, and suggested solutions to, social and environmental problems (Eyles, 1974, 65)

the understanding of the patterns which arise from the use social groups make of space as they see it, and of the processes involved in making and changing such patterns (Jones, 1975, 7)

(it) stressed structure relations in the analysis of social problems ... Analysis (is) based on interrelated material reality and the social contradictions this produces; which are seen as the motive force for change, and thus responsible for the development of problems like different level-of-living conditions (Asheim, 1979, 8)

the study of consumption, whether by individuals or by groups (Johnston, 1981, 205)

it is an interactionist perspective which aims to uncover how social structure is defined and maintained through social interaction, and which studies how social life is constituted geographically through the spatial structure of social relations (Jackson and Smith, 1984, vii)

While it must be recognised that the removal of these definitions from their contexts does their authors a disservice, certain key features are discernible, notably the foci of individual and group lives and creativity and the relations between social geography and the history of geographical ideas in general. Thus, we see in these quotations social geography as the following: regional geography, spatial sociology, group-environment relations (and the Vidalian tradition), welfare geography in a societal context, human agency, structural marxism, distributional welfare, and as interpretative geography (allied to interpretative sociology and anthropology) respectively. This pigeonholing does a greater disservice to the various authors but allows a view of social geography as a context-dependent discipline, mirroring the battlefield of human geography in general. We shall note these changes in the various chapters in this volume, usually in the form of regional description - areal differentiation - person/environment relations and welfare-relevance; radical-structuralist; and humanistic-interpretative perspectives. There is, however, a deafening silence in the definitions, although perhaps a better way of phrasing it would be to note that a certain form of endeavour is so common that it is given a taken-forgranted status. This endeavour contains the perspectives of locational analysis-spatial science, themes which form central elements of the many of the chapters. Again, it may be that this interpretation is unfair, as much of the discussion on the conceptual nature of social geography has been critical of the application of quantitative methods to the study of social concerns and of the models of individuals and of society assumed by such applications. Yet, the practice of social geography as an empirical subject is dominated by an empiricist approach. We thus need a conception of the subject which relates clearly its conceptual frontier and its empirical character.

Diffusion and Convergence?

Such a relationship may be treated as part of the history and development of a science. Kuhn's (1962) ideas of a normal science (a consensus of practice) and occasional revolutionary disruptions has been shown to be a poor representation of geographical change (see Johnston, 1978) while the ideas of Lakatos (1978) on research programmes have more appeal. He suggests that there is a core of beliefs that provides the basis for practice itself and the tests for validity. Several programmes can exist at the same time in one discipline and they compete for support among the community of scholars. Even this scheme appears to give scant justice to the importance of social context. In other words, both Kuhn and Laktos examine the academic context of science and the ways that this enhances or holds back particular lines of enquiry. But as Wolin (1968) suggests with respect to political theory in particular, ideas and scientific practice and change emanate from crises in the world not in the community of scholars. This is a pertinent comment for the practice of the social sciences which are in effect our studies of ourselves. In such a context, the way of looking at and examining the world has moral and political as well as scientific connotations (see Ryan, 1972). Science and society become inextricably interwoven, although such a statement is a long way from claiming that scientific practice is determined by social realities and requirements. A contextual approach (Berdoulay, 1981) which emphasises how a discipline becomes institutionalised and formalised (Capel, 1981) is, however, extremely useful in examining this interrelation. It must allow recognition of specific conditions and events in particular places and at particular times. Thus, we must recognise that social events in particular places and at particular times. Thus, we must recognise that social geography developed in different contexts which had and continue to have different national interests, social practices, scientific practices and educational values. It is not a matter of social geography evolving in one place (though France and North America are important origins of ideas and practices), or of another set of values and practices becoming dominant and leading to disciplinary convergence (though the Anglo-American views of spatial science and philosophical enquiry have had widespread effects through the size of their social geographies, the domination of communication systems by their societies and the tyranny of the English language).

We must of course note the importance of the French tradition and the present-day salience of Anglo-American enquiry, but to speak of diffusion from the French source and convergence around the Anglo-American model is to adopt a naive evolutionary perspective. Not that such a perspective need be naive and such a scheme has already been employed to describe some of the changes in the nature of social geography (Eyles, 1984). A sophisticated evolutionary format allows for the recognition of several points of origin, several diffusion routes and the continued existence of social geographies no longer on the conceptual frontier. Such 'relicts' present a vital corrective to the 'frontiersmen and women'. The 'backwoods' is our heritage, our source of continuity. We should not look upon such social geography as outdated or outmoded (see Ford, 1982). Indeed, in certain places there is a return to the concerns with and of place identified by the earlier foci of France, Britain and the United States. This return does not herald another form of convergence at a more particularistic level. It demonstrates that due consideration is now being given to local conditions, events and processes as in southern and black Africa, South Asia and Australasia. In turn, considerations of the local lead to an awareness of present and future problems faced or to be faced by a society. Indeed, the future of Poland dominates the paper by Jalowiecki, where the close connection between geography and planning appears to result in social geography being identified as a form of critical social planning. But we must recognise the problems with the term 'social' in Eastern Europe and also point to the important studies of residential and socio-economic differentiation carried out in the spatial scientific tradition (see, for example, Barbash in Smith 1979; Weclawowicz, 1979).

Evolution is not unidirectional nor does it occur in a continuous flow. A subject's development can be greatly shaped by new stimuli from disparate sources which appear along the route so to speak. We must note the significance of Swedish social geography in this respect. From this source has emanated time geography as well as the continued significance of the social geography-policy relationship. The importance of policy and decision-making is also found in the establishment of behavioural geography as the major theme in German social geography, coupled with the use of quantitative techniques for modelling the urban environment (see Lichtenberger, 1978).

Context, academic and social, shapes, therefore, the nature of social geography and affects its evolution. But social geographers themselves along with other scientists also 'make' their own subjects. How else could it be? And what else could a specific social geography be but a combination of the international and the local with greater or lesser degrees of each? The picture obtained of trends in a phenomenon depends, as all geographers know, very much upon the scale or level of analysis, with convergence and coherence becoming diversity and particularism at a smaller scale. It may well be now that the converging tendency of quantitative methodology is being loosened and that diversity and particularism will dominate the immediate future of social geography. Attempts are underway to construct 'indigenous' theories to explain better local conditions (see, for example, Rogerson and Browett and Lee in this volume) and the qualitative analyses, as being undertaken, for example, in Britain, North America, Sweden and the Netherlands, demand a localised, small scale focus in practice if not in conceptualisation and theorisation. This is not, however, a parochial future as in the local we may better see and understand the relationship between individuals and society and between space and society (see Eyles, 1986). These relationships are the underlying themes of social geographical endeavour and point to social geography as part of a broad-based social science and to an interesting future. This is not, however, a parochial future as in the local we may better see and understand the relationships between individuals and society and between space and society (see Eyles, 1986). These relationships are the underlying themes of social geographical endeavour and point to social geography as part of a broad-based social science and to an interesting future.

References

Asheim, B.T. 1979 Social geogrpahy - welfare state ideology or critical social science? *Geoforum* 10, 5-18

Berdoulay, V. 1981 The contextual approach, in D.R. Stoddart (ed) *Geography, ideology and social concern*. Blackwell, Oxford

Buttimer, A. 1968 Social geography, in D.L. Sills (ed) *International encyclopaedia of the social sciences* vol. 6. Macmillan, New York

Capel, H.	1981	Institutionalisation of geography and strategies of change, in D.R. Stoddart (ed) Geography, ideology and social concern. Blackwell, Oxford
Corna-Pellegrini, G.	1980	Numero monografico dedicato a 'geografia a percezione dell' ambiente', Rivista Geografica Italiana 87, 1-132
Eyles, J.	1974	Social theory and social geography, Process in Geography 6, 22-87
Eyles, J.	1984	The examination of social relationships in space GeoJournal 9, 247-53
Eyles, J.	1986	The geography of everyday life, in D.Gregory & R. Walford (eds) New horizons in human geography, Macmillan, London
Ford, L.R.	1982	Beware the new geographies, Professional Geographer 34, 131-135
Ginkel, H. van	1983	Mirror of a changing society, Tijdschrfit voor Economische en Sociale Geografie 74, 358-66
Heslinga, N.W.	1983	Between German and French geography, Tijdschrift voor Economische en Sociale Geografie 74, 317-34
Jackson, P. & S. Smith	1984	Exploring social geography. Allen & Unwin, London
Johnston, R.J.	1978	Paradigms and revolutions or evolutions, Progress in Human Geography 2, 189-206
Johnston, R.J.	1981	The state and the study of social geography, in P. Jackson & S. Smith (eds) Social inter-

		action and ethnic segregation. Academic Press, London
Johnston, R.J. & P. Claval	1984	Geography since the second world war. Croom Helm, London
Jones, E.	1975	Introduction, in E. Jones (ed) Readings in social geography. Oxford University Press, London
Kuhn, T.S.	1962	The structure of scientific revolutions. University Press, Chicago
Laan, L. van der & A. Piersma	1982	The image of man: paradigmatic cornerstone in human geography, Annals, Associations of American Geographers 72, 411-26
Lakatos, I.	1978	Falsification and the methodology of scientific research programmmes in J. Worrall and G. Currie (eds) The methodology of scientific research programmes vol. 1. University Press, Cambridge
Lichtenberger, E.	1978	Quantitative geography in the German-speaking countries, Tijdschrift voor Economische en Sociale Geografie 69, 362-73
Paassen, C. van	1976	Human geography in terms of existential anthropology, Tijdschrift voor Economische en Sociale Geografie 67, 324-41
Paassen, C. van	1984	Human geography in the Netherlands, in Johnston and Claval op. cit
Quaini, M.	1974	Marxismo e geografia. La Nuova Italia, Florence
Ryan, A.	1972	'Normal' science or political ideology? in

		P. Laslett et al (eds) <u>Philosophy, politics and society</u>, Fourth Series. Blackwell, Oxford
Smith, D.M.	1979	<u>Where the grass is greener</u>. Penguin, Harmondsworth
Vila-Valenti, J.	1984	The Iberian peninsula and Latin America, in Johnston and Claval <u>op. cit</u>.
Vooys, A.C de	1950	<u>De ontwikkeling der sociale geografie in Nederland</u>. Wolters, Groningen
Watson, J.W.	1957	The sociological aspects of geography, in G. Taylor (ed) <u>Geography in the twentieth century</u>. Methuen, London
Weclawowicz, G.	1979	The structure of socio-economic space in Warsaw in 1931 and 1970, in R.A. French and F.E.I. Hamilton (eds) <u>The socialist city</u>. Wiley, Chichester
Wolin, S.S.	1968	Paradigms and political theories, in P. King & B.C. Parekh (eds) <u>Politics and experi-ence</u>. University Press, Cambridge
Zerbi, M.C.	1984a	Geografia e qualita' della vita: strumenti, metodi e problemi, in G. Schmidt (ed) <u>Gli indicatiro ambientali</u>. FAST, Milan
Zerbi, M.C.	1984b	Qualita' della vita nelle province italiane, in G. Schmidt (ed) <u>Gli indicatori ambientale</u>. FAST, Milan

Chapter One

SOCIAL GEOGRAPHY IN FRANCE

Paul Claval

There is a long tradition in France of curiosity for social geography: the term was coined in the 1890s by Le Play's disciples, and the dominant tradition of vidalian geography was more open to social aspects than was the case with foreign schools. Until the 1950s, there was nevertheless no effort to cover this aspect of the field by theoretical interpretations. During the last 30 years, many avenues have been explored: inspiration came from marxism, from English or American geography, from French or foreign sociology, from anthropology, from history as practised by the Annales school and from the study of ideologies. Progress has been slow until the last ten years. The field is still divided between many orientations: it has become central to an important group of academics during the last few years, and theoretical discussion is by now very active.

Origins: The Vidalian Tradition

The Le Play school provided geography with the term of social geography (Dunbar, 1977), but had no lasting influence on the main currents among French geographers (Berdoulay, 1981). People have to look to Vidal and his disciples to understand the geography which was built between 1890 and 1910 and dominated the academic field until well into the 1950s.

French geographers were trained as historians as well as geographers: it gave them the idea of possible variations in man-milieu relationships - hence the possibilism - and the idea too of the significance of social aspects of life and of the diversity of civilisations. But Vidal de la Blache (1913, 298) did not conceive geography as a social science: for him, it was a "science of places and not a science of men". Geographers preferred to adopt the predominant natural science paradigm.

Social orientations were not completely ignored, but there was no way to explore them directly. It was mainly through the analysis of ways of life (genres de vie) that social considerations were introduced, but not explicitly.

When ways of life are described, three dimensions are - or can be - explored: 1) ways of life are defined by the tools, the techniques and the time used to exploit the environment; 2) ways of life describe the networks of social relations which link individuals and groups, and the time necessary for this type of activity; 3) ways of life appear also as normative models: in some societies, for instance, nomadism is highly valued; elsewhere, peasant farming is preferred.

The first aspect (objective modes of life) offer few opportunities for exploring social facts; the second one is mainly social, and the third one, with its emphasis on styles of life, stresses the difference between objective and subjective aspects of social life. French geographers were mostly interested in the material aspects of ways of life (point 1), but they did not ignore points 2 and 3: they often gave descriptions of social relations and some glimpses of the way people experienced space (through the curiosity for noms de pays (see Sopher, 1974)), and through the discovery, especially in mediterranean countries and in black Africa, of the normative aspect of ways of life (see Labouret, 1941; Weulersse, 1946) - of styles of life. These orientations were particularly conspicuous in the doctoral dissertations of Camille Vallaux (1907) and Jules Sion (1908) - but Vidal de la Blache (1909) considered they had overemphasised the social side of reality.

During the interwar period, regional geography often gave interesting views of social situations in rural areas, but the emphasis was generally on other aspects. Cholley (1931) initiated a very clever cartography of social stratification in rural areas around Paris - but he was not imitated.

In the 1930s, more and more attention was given to rural landscapes, and to the ways agrarian structures were developed: inspiration came from Marc Bloch (1931), Roger Dion (1934) and Gaston Roupnel (1932). As a result, more and more attention was given to the social conditions which were dominant at the time the rural landscapes were organised:

there is a strong social bias in the theses of Maurice Le Lannou (1941), Pierre Brunet (1960) and Etienne Juillard (1953). In Alsatia, where Juillard worked, rural differentiation is linked to the distribution of religious groups and to the size of farms: it is possible to link the history of agricultural progress to the particular social structures and cultural traditions of each local area.

At mid-century, French geography was more open to social problems than other schools, but there is nothing like a recognised social geographical field.

New Curiosities and New Orientations in the 1950s and 1960s

Geographers spent more and more time, during the 1950s and the 1960s, on social characteristics of the areas and of the problems they were studying - but they did this with so diverse motivations that there was no more unified social geography in 1970 than in 1950.

Some geographers thought it necessary to give more coherence to the vidalian tradition, or to build geography on more critical bases. Maurice Le Lannou (1949) was the more faithful to the vidalian tradition, but interpreted it in such a way that his interest was more and more on the way people inhabit their country: his views were conducive to more sensitivity to the experience of place and to the ways people conceive their work. Renée Rochéfort (1963) opened one of the first avenues in social geography by her thesis on Le travail en Sicile, but she remained isolated during the 1960s.

Etienne Juillard developed historical interpretations of the way regions evolved since the end of the XVIII century, and his students showed the role of different social groups in the transformation of territorial organisations: they looked at the decline of local communities and at the emergence of new forms of social relations (Schwab, 1980). But this orientation became popular only in a small group of former students of the University of Strasbourg.

André Cholley (1942, 13) gave more emphasis to social aspects of geography when he defined the discipline as a science in combinaisons, that is in the relations between complex phenomena. "An agrarian combination, for instance, gives to the human group it concerns a particular social structure, with a more or less satisfying income level..." Such a conception is conducive to precise analyses of social organisation.

Max. Sorre was the more open to sociology. In his Fondements de la géographie humaine, he looked often toward sociology for inspiration. This orientation was even more obvious in Rencontres de la géographie et de la sociologie (Sorre, 1957), in which his views on society and space were greatly influenced by the Chicago School and Halbwachs (1938). Anne Buttimer (1971) showed the influence of Max. Sorre's conceptions on the ideas developed by the French sociologist Chombart de Lauwe (1952). Geographers were, at the time, less receptive.

In the 1950s, the new social orientation in human geography was most popular among marxists. Pierre George (1937, 1945, 1966) and Abel Chatelain (1946, 1947) were interested in French social geography and Jean Dresch (1979) explored the geographical implications of imperialist structures in French colonies. With the translation of German ideology, interest grew in the relations between cities and their rural neighbourhoods. Michel Rochefort (1950) was the first to draw attention on the role of capitalism in rural modernisation. Raymond Dugrand (1963) and Roger Brunet (1965) tried to measure the degree of domination of rural areas by urban-based aristocracies or bourgeoisie. By the mid-1960s, these studies appeared at the same time fruitful and deceptive: they required time consuming archival exploration to build accurate pictures of land ownership and social relations and did not offer fresh possibilities of interpretation. Everything was known beforehand: empirical data were needed only to illustrate aspects of marxist theory.

By the end of the 1960s, social geography appeared as a necessary chapter in human geography; many colleagues had tried to structure it, but there was no agreement on its scope, its methods and its role.

The 1970s: In Search of Coherent Frameworks

At the end of the 1960s, French geographers became aware of the new developments in social analysis in English-speaking countries. It was the time when many colleagues had opportunities to teach in Quebec universities: they became acquainted there with modern factorial ecologies. Young colleagues like Antoine Bailly with Mario Polese (1977, 1978) and Marc Bonneville (1978) used them either for Canadian cities or for French ones. But there was soon a feeling of dissatisfaction. Factorial ecologies provided geographers with facts, but not with interpretations. A good deal of reflection was devoted to this problem in the 1970s.

There are two major ways of conceiving social geography: 1) social geography can be conceived as a description of the diversity of social conditions and of their spatial patterns; 2) it can be thought of as the study the way social relations shape human distributions.

The majority of French orientations until the end of the 1960s as well as modern factorial ecologies resort to the first category. What geographes wished mainly in the 1970s was to develop satisfying theoretical interpretations: attention focused on social relations.

Traditional sociology had few models to offer for building spatial theory. The notions of class structure and class relations were good as long as description was the aim, but it was difficult to use them for spatial interpretations. The main reason for that was the lack of precision in the way they were used. Sociologists and geographers were not always attentive to the difference between classes as collectivities of persons with similar objective characterstics, and classes as groups conscious of their identities and interests: since the transition from the first situation to the other involves very important spatial processes (there can be no class consciousness without possibilities of intercommunication), lack of precision in this field precluded all theoretical development (see Claval, 1973).

French academic sociology had a few tools to offer to geographers at the end of the 1960s, but French geographers did not use all of them: they were not aware of the fecundity opened by time perspectives when analysing life cycles. Murdie (1969) had been keener than French academics to exploit Chombart de Lauwe's orientations.

French geographers were more open to marxist interpretations: many of them had been communists in the 1950s and sometimes in the 1960s and were still relying on marxist categories even if they had ceased to act as marxists. At the beginning of the 1970s, marxist sociologists were developing new and exciting analyses of the urbanisation process; they were showing the role of State authority and of State investment in urban development and interpreted it as a sign of ageing capitalism. To avoid profit deterioration, public money was needed.

Henri Lefebvre (1970) was the best known of

these marxist sociologists. French geographers read his books as well as the books and the papers of Lojkine (1973), Préteceille (1973), Rey (1974), Topalov (1974), Lipietz (1977), Castells (1972) and Godard (1974), but their reaction was a cautious one. There was never, among French geographers, the kind of enthusiasm British geographers experienced for Castells' Urban Question. Why? For theoretical reasons, geographers were hesitant when confronted with the new social paradigm. Many were ready to speak of "space production", just like Lefebvre (1974) did, but they were afraid of some of the structuralist interpretations of this conception. Indeed the space we observe around us is not a natural one: it has been transformed through millenia of hunting, gathering, cattle raising and farming. It is really a product of human agency, but it has not been produced in a vacuum. Geographers had the feeling that many sociologists gave so much emphasis to social structures than they ignored the material conditions of social life (see Claval, 1967) - a major contradiction for a marxist interpretation!

The attitude of French geographers dealing with social problems was thus often mainly a critical one: critical of the lack of proper theoretical foundations in factorial ecologies, critical of the structuralist viewpoint of many of the sociologists. As a result, they hesitated and preferred often to avoid the use of criticised methods. Until 1977 or 1978, few empirical studies of urban social situations were published. With firmer grasp over the basic tenets of social analysis of spatial facts, attitudes changed at that time: for six or seven years, more and more geographers have been involved in urban social geography (for a review, see Noin, 1983). There is then a substantial difference with the 1950s and 1960s, when interests had focused mainly on rural-urban social relations.

For geographers with an interest in building a theory of social processes in space - and of resulting territorial organisations - there were other avenues open to investigation.

Why not develop the study of the spatial dimension of social relations? Is not space structured by systems of societal relations, with in-born spatial properties? Is it not possible to explain spatial organisation by the spatial properties of social architecture (see Claval, 1967) - i.e. of the whole fabric of social links present in groups?

French anthropology had developed an interest in

the field of societal relations in the 1960s (Maquet, 1971), and modern sociology offered similar possibilities for modern and industrial societies (Etzioni, 1968). The idea of the range of a good could be extended from the field of economic relations to the field of social relations, becoming a possibility for explaining the spatial properties of each system of social relations. Communication theory provided these new orientations with ideas about information requirements of each system, about the transfer costs they involve and the size of the groups they can unite (see Claval, 1973).

Ways of life are first analysed as collections of roles; the roles are combined by individuals, producing their social personalities. They define, when similar, homogeneous groups, classes or collectivities. When complementary, they are structured in systems of social relations (Claval, 1973).

Social relations have economic, political or social contents: they imply exchanges of goods, of prestige and of orders and influence, and provide social agents with wealth, status or power. The social geography which is built on social relations and their spatial properties is an ambitious one: lato sensu, it covers the half of human geography which is not concerned with man-milieu relationships, but is centred on man to man or group to group links; stricto sensu, it is interested in the part of social networks which rely mainly on prestige distribution mechanisms (Claval, 1973) - economic geography deals with the part of social networks which are built on economic mechanisms; political geography is concerned with the power structures and mechanisms (see Claval, 1978). Similar views on the relationship between social organisation and power are found in Giddens (1981). In modern developed societies, the three sectors are often independent; in traditional societies, social relations involve at the same time economic, political and social mechanisms: it means that it is impossible to disentangle social geography (stricto sensu) from economic and political geography.

The mechanisms analysed in this way by social geography are not independent of the conceptions people have of their conditions and of their life. It means that ideologies are important for understanding the spatial properties of social systems, and that social geography has close links with cultural geography (Claval, 1984).

This theoretical interpretation of social geography has been influential in France for ten years, but few geographers have endorsed it completely: they lack the theoretical orientation it needs, and they regret its liberal foundations.

French sociology has changed during the last ten years. It is impossible to understand the present state of social geography without a view of this evolution and its influence among geographers.

The study of peasant societies began to be popular in France at tne end of the 1950s. Until the end of the 1960s, the work by Mendras (1967), Rambaud (1969), or by Duplex (1968) had little influence on geographers. They considered that it was important to look at the disappearance of the former peasant society, and at the way it has been replaced by new forms of peasantry - since it is one of the most original theses of Henry Mendras (1980, 15-17): the French countryside is still different from heavily urbanised areas because people have developed different forms of social ties, of social consciousness, and rely on different economic mechanisms. Robert Chapuis (1981), for instance, is using sociological techniques to build a view of geography which is clearly sociological in its orientation.

A better view of the social and political mechanisms and of their spatial properties had been provided by Norbert Elias (1974) and by Pierre Gremion (1976). Norbert Elias' study of the social and cultural functioning of the <u>société de cour</u> as developed by French absolutist monarchy was written more than forty years ago, but it was translated into French about ten years ago. It was important for geographers. Generally, they think that purely social mechanisms, that is mechanisms based on prestige competition, are efficient only in small confined societies, since the range of prestige relations is a short one: individuals have to see the people who praise them to measure their sincerity and to enjoy their behaviour. What Louis XIV and other monarchs had discovered was the way to use prestige competition to rule over a large kingdom: nobility had to spend several months each year at the court; it was necessary for all aristocrats if they wished to succeed in the prestige competition it being so important to get the king's favour. As for all prestige competition, the process could work only in a very restricted area. But it took a national dimension since aristocrats owed the political influence they exerted in their provinces to the Versailles process.

Pierre Grémion (1976) has an interest in

present-day political processes. He has been fascinated by one of the major paradoxes of French society: the political system has been heavily centralised since the seventeenth century, and French society is more diverse than any of the national societies in Western Europe. Why? Because centralisation was not devised to erase differences, but to give the power to decide to national authorities, i.e. decisions with local impacts have to be taken. But they are formally proclaimed in Paris, being prepared in provincial cities by civil servants and local notabilities. Both groups are interested in avoiding local conflicts, hence the lasting diversity. Such social and political studies are important to enlighten the complexity of spatial dimensions of social processes.

French spatial sociology is becoming more systematic. For many young sociologists, society is conceived as a stage: they took from Goffman (1973) the idea of major oppositions between what is going on the stage, and what is accomplished behind the stage. Louis Marin (1973) and Pierre Sansot (1978) systematised this idea when they began to speak of primary and secondary spaces - the space of the second home can serve as a paradigm for secondary space. Michel Foucault's (1975) emphasis on supervision (surveillance) is important in underlining the importance of what is going on on the state for social control.

Jean Remy and Liliane Voyo (1981) used these ideas to show the ambiguity of places in social life: cities are at the same time places for order and places for violence, and both forms have to be experienced at the same time if cities have to take significance in human life.

The experience of space is often enshrined in stereotyped patterns: Pierre Bourdieu's (1979) idea of habitus may be useful to explain the genesis of these patterns. They define - using Bourdieu's categories - fields for action, and these fields may be defined as categories of discourse. Attitudes towards central cities, for instance, have changed consistently since the beginning of the 1960s with the emergence of new social values, new preferences; out of them, norms for restoring the old parts of the city have been developed, social agents have developed their roles; production of space ceases to appear as the result of impersonal structures.

Alain Bourdin (1984) is one of the sociologists who are elaborating this new spatial sociology. He is offering a more systematic framework to the findings of research developed since the end of the 1960s in Toulouse, in the group of urban sociology headed by Raymond Ledrut (1968, 1973, 1976). Studies like the analysis of social life in French small cities by Jean-Claud Lugan (1983) could have been produced by geographers. Xavier Piolle (1979), in Pau, has developed research on lines parallel to those of Ledrut.

As a result, more attention is now devoted to problems of life experience: geographers look at the way different groups build their action space and experience it. The inspiration came first from Chombart de Lauwe and from the study of mentalities by historians of the Annales school. Some roots lie also in the old analysis of choronyms: Jean Gallais (1967) developed his ideas about social experience of different groups in the tropical environment using this idea. Armand Frémont (1981, 1983) and his colleagues in Caen (CERA, 1977) rely more on literary sources and on the Annales school. Gildas Simon (1979), in his studies on social determination of migrations, was more open to sociological models, while many of these approaches were combined by Vant (1981) on images and the urbanisation process.

Present Situation and Perspectives

There is more and more interest in social geography: during the 1970s, many geographers were trying to build something new in this field, but they hesitated between different orientations. The output was not very important, even if many people were involved in the field.

The situation has changed in the last two or three years. A working group has been organised within the French national committee (see Noin, 1983). It links people working in Caen, in Paris, and in Lyon, the three main centres for social geographers in France. There are minor groups elsewhere, in Strasbourg, working along Juillard's lines, in Toulouse, with a strong marxist bias, but with close contacts with urban and rural sociologists in Pau and in Nantes. Indeed, social geography is now developing sufficiently to attract other views of its nature (see Herin, 1982; Reynard, 1982).

Social segregation in cities, ghetto formation and differential experiences of space by different social groups or classes will certainly play a dominant role among empirical studies in the next few years . Studies on these themes have been carried out by Brun and Chauvire (1983), Guibourdenche and Joly (1979), Joly (1978), Parlhe (1978), Roncayolo (1972), George (1978) and Noin (1980). In the theoretical field, some syntheses of the approaches imagined in the 1970s will be certainly elaborated.

References

Bailly, A. & M. Polese	1977	Analyse typologique en milieu urbain: les aires homogènes d'Edmonton, L'Espace Géographique, 4, 267-275.
Bailly, A. & M. Polese	1977	1978 La différentiation de l'espace urbain de Québec, Cahiers de Géographie de Québec, 22 , 22-49
Berdoulay, V.	1981	La formation de l'école francaise de géographie (1870-1914), Paris, Bibliothèque Nationale
Bloch, M.	1931	Les caractères originaux de l'histoire rurale francaise, Oslo, Institut pour l'étude comparée des civilisations
Bonneville, M.	1978	Naissance et métamorphose d'une banlieue ouvrière, Lyon, Presses de l'Université de Lyon
Bourdin, A.	1984	Le patrimoine réinventé, Paris, PUF
Brun, J. & Y. Chauvire	1983	La ségrégation sociale, in Noin op. cit
Brunet, P.	1960	Structure agraire et économie rurale des plateaux a teritares entre la Seine et l'Oise. Caen, Caron
Brunet, R.	1965	Les campagnes toulousaines. Etude géographique, Toulouse, Boisseau
Buttimer, A.	1971	Society and milieu in the French geographic tradition, Chicago, Rand McNally

Castells, M.	1972	La question urbaine, Paris, Maspéro
Castells, M.	1974	Monopolville, Paris, Mouton
Centres D'Etudes Regionales et D'Amenagement (CERA)	1977	Stuctures sociales et espace vécu dans le Bocage Normand, Caen, University of Caen
Chapuis, R.	1982	Les ruraux de départment du Doubs. Eléments de géographie sociologique, Besancon, Cêntre
Chatelain, A.	1946	Cette nouvelle venue, la géographie sociale, Annales, Economies, Sociétés, Civilisations, 1, 266-270
Chatelain, A.	1947	Les fondements d'une géographie sociale de la bourgeoisie, Annales de Géographie, 56, 455-462
Cholley, A.	1931	Essai d'une carte de représentation de l'habitat rural, C.R. du Congrès International de Géeographie, Paris,
Cholley, A.	1942	Guide de l'étudiant en géographie, Paris, PUF
Chombart De Lauwe, P-H	1952	Paris et l'agglomération parisienne, Paris, PUF, 2 vol
Claval, P.	1967	Géographie et profondeur sociale, Annales, Economies, Sociétés, Civilisations, 22, 1005-1046
Claval, P.	1973	Principes de géographie sociale, Paris, Thérèse Genin et Litec
Claval, P.	1977	Le marxisme et l'espace, L'Espace Géographique, 6, 145-164
Claval, P.	1978	Espace et pouvoir, Paris, PUF
Claval, P.	1984	La géographie humaine et économique, Paris,

PUF

Dion, R. 1934 *Essai sur la formation du paysage rural francais*, Tours, Arrault

Dresch, J. 1979 *Un géographe au déclin des Empires*, Paris Maspéro

Dugrand, R. 1963 *Villes et campagnes du Bas-Languedoc*, Paris, PUF,

Dunbar, G. 1977 Some early occurrences of the term "Social Geography", *Scottish Geographical Magazine*, 93, 15-20

Duplex, J. 1968 *Atlas sociologique de la France rurale*, Paris, A. Colin

Elias, N. 1974 *La société de cour*, Paris, Calmann-Lévy

Etzioni, A. 1968 *The active society. A theory of societal and political processes*, New York, Free Press

Foucault, M. 1975 *Surveiller et punir*, Paris, Gallimard

Fremont, A. 1981 Une Problématique de la géographie sociale, Caen, *Bulletin du CERA*, 12, 5-8

Fremont, A. 1981 La classe ouvrière dans la France de l'Ouest. Esquisse de géographie sociale, in Noin *op. cit*.

Gallais, J. 1967 *Le delta intérieur du Niger. Etude de géographie régionale*, Dakar, IFSN, 2 vol

George, P. 1937 *Géographie économique et sociale de la France*, Paris, Hier et Aujourd'hui

George, P. 1937 *Géographie sociale du monde*, Paris, PUF

George, P. 1945 *Géographie sociale du monde*, Paris, PUF

George, P. 1966 *Sociologie et géographie*, Paris, PUF

George P. (ed) 1978 Occupation sociale de

		l'espace urbain, *Bulletin de l'Association des Géographes Francais*, no 449, p. 3-40
Giddens, A.	1981	*A contemporary critique of historical materialism*, vol. 1, *Power, property and the state*, London, Macmillan
Goffman, E.	1973	*La mise en scène de la vie quotidienne*, Paris, Editions de Minuit
Gremion, P.	1976	*Le pouvoir périphérique*, Paris, Le Seuil
Guibourdenche, H. & J. Joly	1979	Changement social et structures spatiales dans l'agglomération et les quartiers de Grenoble, *Revue de Géographie Alpine*, 66, 257-279 and 391-406
Halbwachs, M.	1938	*Morphologie sociale*, Paris, A. Colin
Herin, R.	1982	*Herencias y perspectivas en la geografia social francesa*, *Geocritica*, 42
Joly, J.	1978	Structures sociales de l'agglomération et des quartiers de Grenoble, *Revenue de Géographie Alpine*, 65, 385-407 and 66, 5-28
Juillard, E.	1953	*La vie rurale dans la plaine de Basse Alsace*, Paris, Les Belles Lettres
Juillard, E.	1974	*La région. Contribution à une géographie générale des espaces régionaux*, Paris, Ophrys
Labouret, H.	1941	*Paysans noirs de'Afrique occidentale*, Paris, Gallimard
Ledrut, R.	1968	*Sociologie urbaine*, Paris. PUF
Ledrut, R.	1973	*L'espace social de la ville*, Paris, Anthropos
Ledrut, R.	1976	*L'espace en question*,

		Paris, Anthropos
Lefebvre, H.	1970	*Du rural à l'urbaine*, Paris, Anthropos
Lefebvre, H.	1974	*La production d'espace*, Paris, Anthropos
Le Lannou, M.	1941	*Pâtres et paysans de la Sardaigne*, Tours, Arrault
Le Lannou, M.	1941	*La géographie humaine*, Paris, Flammarion
Lipietz, S.	1977	*Le capital et son espace*, Paris, Maspéro
Lojkine, J.	1973	*La politique urbaine dans la région parisienne*, Paris, Mouton
Lugan, J-C	1983	*La petite ville au présent et au futur*, Paris, CNRS
Maquet, J.	1971	*Pouvoir et société en Afrique*, Paris, Hachette
Marin, L.	1973	*Utopiques. Jeux d'espaces*, Paris, Editions de Minuit
Mendras, H.	1967	*La fin des paysans*, Paris, SEDES
Mendras, H.	1980	Une figure de la France, in, H. Mendras (ed.), *La sagesse et le désordre. France 1980*, Paris, Gallimard, 15-71
Murdie, R.A.	1969	*Factorial ecology of metropolitan Toronto, 1951-1961; an essay on the social geography of the city*, Chicago, Department of Geography, Research Paper 116
Noin, D. (ed)	1980	Séance sur la géographie sociale de l'agglomération parisienne, *Bulletin de l'Association des Géographes francais*, 466-467, 50-87
Noin, D. (ed)	1983	*Géographie sociale. Actes du colloque de Lyon, 14-16 Octobre 1982*, Paris, Guepes and

University of Paris-I

Pailhe, J.	1978	Les transformations de la composition sociale de la ville de Bordeaux, Revue Géographie des Pyrénées et du Sud-Ouest, 49, 11-28
Piolle, X.	1979	Les citadins et leur ville, Toulouse, Privat,
Preteceille, E.	1973	La production des grands ensembles, Paris, Mouton
Rambaud, P.	1969	Société rurale et urbanisation, Paris, Le Seuil
Remy, J. & L. Voye	1981	Ville, ordre, violence, Paris, PUF
Rey, P.P.	1974	Les alliances de classe, Paris, Maspéro
Reynaud, A.	1982	La géographie, science sociale, Reims, Tigre
Rochefort, M.	1950	La pénétration des capitaux bourgeois dans la campagne autunoise, ses conséquences sur l'habitat et la structure agraire, Revue de Géographie de Lyon, 25, 249-266
Rochefort, R.	1963	Travail et travailleurs en Sicile, Etude de la géographie sociale, Paris, PUF
Roncayolo, M.	1972	La division sociale de l'espace urbain, Bulletin de l'Association des Géographes Francais, 395-396, 5-20
Roupnel, G.	1932	Histoire de la campagne francaise, Paris, Grasset
Sansot, P. et al	1978	L'espace et son double, Paris, Editions du Champ urbain
Schwab, R.	1980	De la cellule rurale à la région. L'Alsace 1825-1960, Paris, Ophrys
Simon, G.	1979	L'espace des travail-

		leurs tunisiens en France, Poitiers, Simon
Sion, J.	1908	Les paysans de Normandie orientale, Paris, A. Colin
Sopher, D.	1974	Toward a rediscovery of India: thoughts on some neglected geography, in M. Mikesell (ed), Geographers abroad, Chicago, University of Chicago, Department of Geography Research Paper, 110-133
Sorre, M.	1943-53	Les fondements de la géographie humaine, Paris, A. Colin, 4 vol.
Sorre, M.	1957	Rencontres de la géographie et de la sociologie, Paris, Marcel Rivière
Topalov, C.	1974	Les promoteurs immobiliers, Paris, Moutono
Vallaux, C.	1907	La Basse-Bretagne, étude de géographie humaine, Paris, PUF
Vant, A.	1981	Imagerie et urbanisation. Recherches sur l'exemple stéphanois, Saint-Etionne, Centre d'Etudes Foreziennes
Vidal De La Blache, P.	1909	Les paysans de Normandie orientale, Annales de Géographie, 18, 177-181
Vidal De La Blache, P.	1913	Des caractères distinctifs de la géographie, Annales de Géographie, 22, p. 289-299
Weulersse, J.	1946	Paysans du proche-Orient, Paris, Galimard

Chapter Two

NORTH AMERICA

R.J. Johnston

Social geography has never occupied a central place within the discipline in North America, as will be illustrated here. The main academic traditions in the period leading up to the Second World War focused on either the mapping of economic artifacts in the landscape (notably land use) or the study of landscapes as reflections of culture (see Porter, 1978). In neither of these did social life and individual agency form substantial foci. In the 1950s and 1960s some attention was directed towards the mapping of social phenomena, largely in terms of residential patterns, but this was mainly as a sub-section of urban geography and no clearly-defined subdiscipline of social geography, with clear links to the non-economic social sciences, was identified. More recently, attempts have been made to develop work in social geography, in two separate contexts - Marxist-inspired analyses which see social and economic phenomena as inextricably inter-twined, and humanistic analyses.

A Weak Base

The absence of any substantial foundation of social geography in North America is readily documented with reference to a small number of influential books. The first of these was Hartshorne's The nature of Geography (1939), which became a major statement of orthodox geographical philosophy and methodology in mid-century. Social geography receives only two references in the index: in one case, it was used as a synonym for human geography - countering physical (p. 44); in the other, concerned with the study of political boundaries, Hartshorne refers to 'various other topics in what may be called "social geography"' (p. 249).

Hartshorne also indexes 'sociological geography', which he defines as a 'theoretically possible field' (p. 404), linked to what is more generally known as cultural geography. Thus, for example he writes that

> it may appear doubtful that the development of studies of that character can make important contributions to geography. In areas of primitive development one might study the geography of clothes, or implements, or conceivably of manners and customs, and religions. For the important parts of the world, however, such studies would apparently have little geographic significance (p. 405).

Further, in his only detailed reference to the discipline of sociology, he argues against the view that the study of nonmaterial phenomena belongs to sociology, not geography (p. 214), but argues that

> if nonmaterial phenomena are going to be needed ultimately in a regional study, they had better be considered as they are encountered in the field (p. 215).

Twenty-years later, in a volume designed to update the earlier one - Perspective on the Nature of Geography (1959) - Hartshorne had no index entry for social geography. Sociology received one mention, in a discussion of quantification and law development. Perhaps it is little wonder that, even twenty years after that, when Wilbur Zelinsky wanted a chapter on social geography for his volume on 'Human geography: coming of age', he turned to two British authors (Eyles and Smith, 1978).

The landmark volume American Geography: Inventory and Prospect (James and Jones, 1954) also makes clear the virtual absence of social geography from North American geography at that time. None of its twenty-six chapters deals with social geography - nor, much more surprisingly, is there one on cultural geography (most of the references to Sauer are in the chapter on historical geography, by one of Sauer's students, Clark). Social geography is not in the index either. Sociology and its relevance to geography is present, but the only substantial reference is to the area of study known as human ecology. Whittlesey (1954) notes that the term was coined by Barrows (1923) to cover the relationships between people and places, but that sociologists use it to refer to the symbiosis within human communities and are at pains to differentiate it from 'geography as human ecology'. Further, he notes that even those who embrace both people-place and people-people interrelationships as the basis for human ecology tend to ignore the place-place links, but

> These ... ecological aspects of the natural sciences ... were early formulated and long since firmly established. Only a very slight difference of viewpoint distinguishes an ecological study from a geographical study in a typical field. But since that difference concerns space, the basic organising concept of the geographer, it is to him all-important (p. 28).

Whittlesey, clearly, viewed human ecology as a synonym for classicial regional geography.

To American geographers before 1960, social geography was very largely terra incognita, therefore - as made abundantly clear in its absence from James and Martin's history of the discipline. It may be, however, that this conclusion reflects a terminological problem, and that social concerns were widely discussed under other titles. Analysis of the literature indicates that this was not so. In cultural geography, for example, the focus has been on artifacts and categories (e.g. religion) rather than on everyday social life and human agency (e.g. Zelinsky, 1973); much of historical geography shares this emphasis. As Duncan (1980) makes clear in a detailed analysis, culture has been reified, given an existence of it own rather than being represented as a human creation, continually reproduced by individual agency; it lacks any sociological content, and contains 'landscape, artifactual and aesthetic emphases' (p. 198). Similarly with population geography, the focus is on aggregates and reified determinants of individual action (e.g. Zelinsky, 1966).

Only in urban geography, itself an under-developed area until the 1960s, is there some evidence of a 'social focus'. Mayer's (1954) chapter in American Geography: Inventory and Prospect emphasises morphological, historical and economic studies, but does include a section on 'internal forms and patterns' which considers social as well as economic and artifactual variables. His main sources are the models developed by Burgess and Hoyt, and he neither quotes geographical studies set in those contexts nor investigates the sociological base for such models developed by the Chicago school. His later collection of readings (Mayer and Kohn, 1959), and his introduction to it, epitomises the lowly status of social geography within the urban context.

A Pioneer Claim

It was in the field of urban geography, and in a competitor to the James and Jones (1954) volume, that the first North American call for a social geography was set. (The author, J. Wreford Watson, is a Scot, but was working in Canada at the time. Throughout this chapter, authors working in North America have been characterised as North Americans - at least while working there.) Entitled 'The sociological aspects of geography', the chapter (Watson, 1953) begins by noting that geographers have always been interested in society - albeit often focusing on esoteric rather than day-to-day elements of social life - creating

> the vital tradition of differentiating regions as much by their social content as by their physical character (p. 463).

However, geographers were trapped into making extravagent claims about the links between physical and social geography (i.e. environmental determinism), and the latter lost favour.

Watson argues for social geography on six grounds: 1) that it is the human element which distinguished places and gives them character, not the physical; 2) that social variables are necessary correctives to the dominance of physical features in the definition of so-called 'natural regions'; 3) that the cultural landscape reflects individual and social reactions to environments; 4) that human regions should be defined in terms of 'values' as well as artifacts; 5) that functional regions involve social as well as economic interactions; and 6) that 'natural areas' defined by sociologists have much in common with geographical regions.

From this, Watson moves towards defining social geography. He proceeds by eliminating four possible definitions, as necessary but insufficient. Thus, for example, since geography 'is not the mere account of distributions on the earth's surface' social geography is not simply the location of the territorial boundaries of social systems. Secondly, he argues that social geography is not only the study of social interaction constrained by territorial boundaries, for this would be no more than the sociology of areas nor is it the study of the geographical basis of society, with the potential perils of environmental determinism, nor simple the regional differentiation of social characteristics. Watson claims that social geography combines all of these, to produce

> the identification of different regions of the earth's surface according to associations of social phenomena related to the total environment (p. 482)

thereby embracing both form and process.

Watson's social geography reflected the strength of the regional geography paradigm, and translated it from a concern with artifacts only to one which considered social as well as material phenomena. He illustrated it with a case study of Hamilton, Ontario, much of which was typical of the sort of regional studies then popular (beginning with the physical environment and moving on to consider historical geography). He extended the traditional approach, however, by mapping social phenomena and institutions. His map of the city's social regions, for example, included a shatter belt (unstable working class), stable working class, unstable better class, and stable better class, as well as industrial and commercial areas, and he looked at the locations of public relief agencies and of various churches. Indeed, one of his concluding paragraphs uses terminology which suggests the sorts of work that only become popular in the 1980s:

> a region may interpret its physical geography quite differently as its social geography changes. _Its character as a region depends not on the influence of natural material force, in the environment, but on social forces which readapt and reorganize the area and give it a unique character_ (his emphasis, pp. 495-496).

(Note, however, that he reifies the 'region').

Geographers Discover Sociologists

Watson's essay had very little influence on geographical work in the 1950s, and there was neither any indigenous development nor much recognition of the important potential links with developments in sociology. Geographers remained in a disciplinary cocoon and largely restricted themselves to their traditional concerns with artifacts and regions.

The situation changed at the end of the 1960s. By then, much else had changed in American geography, with the impact of the so-called 'theoretical and quantitative' revolution (Johnston, 1983). For its first few years, this was concerned with topics that fell outside the ambit of social geography as defined by Watson; indeed Berry (1964a) attempted to justify it to the 'non-believers' as no more than a better methodology for traditional geographical concerns. Later, however, the potential of work in sociology was realised, and social geography began to develop - very largely within urban geography and stressing the spatial science approach (Pooler, 1977).

The Chicago School

Although models such as those of Burgess and Hoyt had been appraised by geographers in the 1940s (Harris and Ullman, 1945) and 1950s (Mayer, 1954) they had no appreciable impact until the 1960s (Murphy's, 1966, text, for example has a chapter - 'Theoretical explanations of city structure' - based on those models, which refers to only two pieces of geographical writing. A preceding chapter on 'People in the city' is concerned only with mapping population distributions). At this time, however, Brian Berry appreciated the utility of multivariate statistical analyses of census small-area data for 'testing' the Burgess *et al.* models; his research agenda for the frontier in urban geography (Berry, 1965) included the statement that

> A real need remains for carefully designed work that will codify, in a series of comprehensive comparative studies, what in fact does exist insofar as the residential pattern of cities is concerned (p. 419).

The initial stimuli for Berry appear to have been the social area analysis of Shevky and Bell and the work on population densities by Clark and others. Regarding the former, Berry (1964b) identified them as providing a 'succinct description of the location of residential areas' (p. 126) and sought to extend the methodology, via factor analysis, described - in telling terms - as

> the new empiricism of the decade, stimulated by advancing computer technology and consequent diffusion of multivariate analysis throughout the social sciences (p. 126).

It allowed him to set the study of urban residential patterns, through the analysis of census data, within his general programme (Berry, 1964a) for reformulating the structure of geographic enquiry thus

> their (the social area analysts') work has now laid the bases for a spatial model of the internal socio-economic pattern of cities in which the relevant role of the traditional concepts is clear (p. 129).

Regarding the latter, the excitement clearly was that of discovering an inductive generalisation - 'Regardless of time or place, this is the pattern to be found' (Berry, 1965, p.419).

This twin focus was carried forward in a range of research studies by Berry and his students. Much of it is summarised in the chapters by Rees (1970) in a text edited by Berry and Horton; the technical developments are considered in another edited volume (Berry, 1971); the spatial elements are analysed in detail by Murdie (1969, 1976); and the approach was extended to historical studies (Goheen, 1970). Urban social geography at this time was almost coincident with the technicalities of factorial ecology.

But No Sociology

The type of work discussed above, which rapidly attracted a large number of similar investigations (many of which are summarised in Davies, 1984) abstracted only a small element of the relevant sociological work and focused on the technical aspects of defining 'social regions' in the city and their spatial arrangement. In terms of Watson's programme for urban social geography they represented the human element in the creation of regions solely as population aggregates.

The links between this work and the human ecology perspective of the Chicago school of sociologists were weak. Robert Park's well-known dictum relating social and spatial distance was accepted as the basic hypothesis to be tested (although factorial ecology does not, when based on correlation coefficients, identify the absolute degree of social area differentiations), as were the models of Burgess and Hoyt relating to spatial arrangements. The richness of the perspective was overlooked, only to be discovered later (and not in North American geography: see Robson, 1969, on the ecological perspective and Jackson and Smith, 1981, on the humanistic elements of Park's work). Park himself, as Entrikin (1980) has pointed out, had a narrow conception of the role of geography within the social sciences, but this did not influence the geographers of the 1960s any more than did the great range of studies of local areas within the city produced by his associates and successors (e.g. Wirth and Suttles).

The views of the Chicago school were not the only approach to human ecology prevalent in sociology in recent decades. Another major set focused on Amos Hawley. His work, especially that of his associates Dudley Duncan and Leo Schnore, was recognised by geographers, but again they concentrated on the technical aspects (e.g. Duncan and Duncan, 1955, on the index of dissimilarity) and the spatial regularities (e.g. Schnore, 1965, on the Burgess model). The theoretical underpinning was largely ignored by geographers (e.g. Hawley's own, 1950, text and an essay by Duncan and Schnore, 1959).

This latter geographical myopia took place despite the attempt by Schnore (1961) to make contacts with geographers. He defined human ecology as the study of social organisation and geography as the study of areal differentiation, and identified clear potential links between the two. The links were not explored by geographers during the 1960s, however, and while quantitative description was rampant, theoretical understanding was slight.

Moving Onwards

A field of study which focuses on description alone is likely to run out of steam fairly rapidly, because the intellectual excitement of new discoveries is no longer present. Thus the study of urban social areas soon became little more than replication in other cities or with new data sets (a further census offered fertile ground) and the only focus of debate was on relatively minor technical points.

During the 1970s, however, the field of urban social geography (although never explicitly defined) expanded its interests in two directions. The first of these involved work on the spatial behaviour of individuals within the context of the regional mosaic of social areas. This work, which rapidly became known as behavioural geography, was stimulated by disenchantment with the relevance of models of spatial organisation then in vogue in economic geography. It turned to social psychology as a source of ideas for studying individual behaviour within spatial structures (a seminal, early collection of papers was Cox and Golledge, 1969), and its implicit goal was the development of inductive generalisations. In the context of the present discussion, the major substantive focus was intra-urban migration, with work on, for example, the spatial pattern of migration as it related to both spatial constraints (distance, direction) and the social area type of origin and destination, on the means whereby people learn about space and spatially-distributed opportunities, and on the role of key individuals and institutions (e.g. realtors) as agents of social area change.

This behavioural work (for a wide-ranging review, see Golledge, Brown and Williamson, 1972) focused on movements between places, therefore, both those which maintained the residential structure and those which led to alterations in the characteristics of individual parts of a social area mosaic. The danger of such studies was that they might be little more than empiricist, 'one-off' descriptions of a particular situation with no wider relevance. This was made explicit early on by Rushton (1969, 392), who distinguished between behaviour in space - 'description of the actual spatial choices made in a particular system' - and spatial behaviour, defined as 'the rules by which alternative locations are evaluated and choices consequently made'. His goal was clearly understanding the latter, which was coupled with a firm belief that underlying rules of spatial behaviour do exist - 'a stability of spatial preferences' (p. 400) - knowledge of which allows both prediction of how individuals will behave in a particular situation and provision of information that allows 'entrepreneurs to change those preferences'. Whether or not most authors accepted those premises, they generally failed to operate within them, and most research reports were empiricist and of little wider substantive relevance.

Linked to this work on flows of people were those of flows of information, organised either to test Hagerstrand's classic ideas regarding spatial biases in the diffusion of information or to assess the impact of local social milieu on individual's attitudes and behaviour. In the former category, the continuing work of Brown (1968, 1982; see also Brown and Moore, 1969) has been seminal, whereas in the latter the suggestions by Cox (1969) regarding spatial patterns in voting studies were influential.

The second direction of developments in the early 1970s represented an extension of the subject matter of such work. Two sub-types can be identified. In the first, the focus was on a wider range of variables than typically employed in factorial ecologies. The lead in this was taken by Smith (1973), in a seminal volume on The Geography of Social Well-Being in the United States. His work was set in the context of attempts to identify indicators of the 'quality of life' at the national scale, which were then very much in vogue. This 'social indicators movement' largely ignored spatial variations in the 'quality of life', he claimed, despite the clear recognition that 'social problems, like economic problems, have an important spatial component, and that their incidence can be subject to extreme areal variations' (p. 3). However, geographers themselves had traditionally ignored this important aspect of social reality, choosing to focus instead on census data which in general pay scant regard to indices of social well-being.

Smith's work was firmly set within the descriptive, regional mould of factorial ecologies, and he used the same set of technical procedures. The difference was in the selection of variables, with Smith focusing on those which represented seven general criteria for defining social well-being (income, wealth, and employment; the living environment; health; education; social disorganization; social belonging; and recreation and leisure). The results of his work were recognised as descriptive only, within the 'general working hypothesis ... that social well-being in the United States is subject to measurable differences from place to place and to fairly regular spatial variations' (p. 7). But he recognised that this mapping was but an initial task, which would both 'sensitise people to the magnitude of the spatial inequities within contemporary society' (p. 137) and inform discussions of how society might be restructured to remove the gross injustices that he described. In these ways, according to Smith,

> professional geography has a vital and revolutionary role to play in the creation of the new society (p. 141).

The general theme was widely accepted, and geographers became involved in both describing spatial variation in well-being and developing explanatory accounts of their origins (e.g. Cox, 1980). Smith's call for their involvement in the 'creation of a new society' became part of a wider debate on the philosophy of geography (see below).

In the other sub-type, the focus was on individual elements in the geography of social well-being. It is exemplified by a series of texts in the McGraw Hill Problems Series in Geography, in which Smith's book was published; it included books on poverty (Morrill and Wohlenberg, 1971), crime and justice (Harries, 1974), health care delivery (Shannon and Dever, 1974), the black ghetto (Rose, 1971), atmospheric pollution (Bach, 1971), and the allocation of public goods (Cox, 1973). Most of these focused on describing spatial variations, although there were attempts to understand their origins (notably in Cox's book) and to suggest how the variations could be alleviated by public policies. A good example of research in the latter context was a large study of the health care system in metropolitan Chicago (summarised in de Vise, 1971), which led to the development of predictive models for the optimal location of hospital and other facilities (e.g. Pyle, 1971).

Not surprisingly at that time, the black ghetto in American cities was a focus of considerable attention, with geographers concerned to contribute to an understanding of the riots and other problems of the late 1960s. Their studies (e.g. Adams, 1972) remained very much in the quantitative-descriptive tradition, however. Morrill (1965) often-cited paper 'The negro ghetto' has as its central feature a probabilistic model of ghetto spread and, despite the general introduction and conclusion, it is hard to find evidence of the commitment that he later wrote of -

> three years of intense involvement in CORE

> on and off the picket lines, obviously influenced and enriched the study beyond that which a totally detailed analysis would have been (Morrill, 1984, 69).

Similarly, Rose's (1970) work had the same 'analytical' focus. Historical scholarship attempted to put the development of immigrant residential areas into a wider context (notably by Ward, 1971; for a review, see Radford, 1981).

Social Geography in the Early 1970s: An Implicit Definition

Although there is little explicit reference to it, Watson's (1953) paper, and the definition of social geography therein, was reflected in the development of the sub-discipline over the next 15-20 years, at least in the emphasis on spatial form. Social geography was predominantly urban in its focus, was particularly concerned with the definition of regions, and drew little - apart from technical material - from related social sciences. Thus Murdie (1969; following Buttimer, 1968)defined social geography as the 'geographical study of social space' (p. 11), with the latter comprising three ingredients:

1. Formal characteristics, summarised spatially as areas and obtained by mapping socio-economic criteria.

2. Functional characteristics, summarised spatially as points which serve as major foci of social activitiy.

3. Circulatory characteristics summarised spatially as lines representing the flows of goods, services, people and ideas and including space perceived by individuals and social groups.

The similarities between this statement and Watson's are many, though Murdie paid relatively little attention to the third ingredient.

Although Murdie adopted Buttimer's definition, it should be noted that the type of social geography reflected in his work (as reviewed above) was but a partial representation of Buttimer's views - themselves strongly influenced by her detailed study (Buttimer, 1971) of French social geography in the

Vidalian tradition. (American sociologists were also influenced by this school - see, for example, Duncan and Pfautz's translation of Morphologie Sociale, 1960 - but geographers overlooked this point of contact.) Buttimer's interests extended beyond the 'objective mapping' of social areas and related phenomena, and she suggested the need for geographers to address the question (Buttimer, 1969, p. 417)

> should we be satisfied with drafting an opaque, objective map of social patterns in space, or must we supplement this with the subjective or inside view?

The answer to this, she suggested, could be found in the works of the French sociologist, Paul-Henri Chombart de Lauwe, with their focus on the 'complex fibre' that holds together a local cultural milieu and on the values, attitudes and preferences of the residents of various social areas.

This approach, if it had been adopted, could have integrated social and cultural perspectives on urban life. In general it was ignored, however, and American social geographers preferred to concentrate on drafting 'opaque, objective maps'. (Despite, it seems, claims that they were irrelevant to the social organisation of city life: Palm, 1973.)

Recent Shifts - In Opposing Directions

Most developments within a social science, after the first flush of enthusiasm, settle down into what Kuhn (1962) has termed 'normal science'. The direction of research has been set, and the procedures to be used defined. Much remains to be done in terms of application and fine-tuning of the procedures, but the likelihood of further very orginal findings becomes increasingly remote. As a consequence, scholars are less attracted to a particular field - although some remain to concentrate on the 'nuts and bolts issues' (see Davies, 1984). Instead, they are more likely to reflect on the achievements to date - to ask what has been learned, and what important issues remain to be tackled.

This has certainly been the case with the field of urban social geography since the early 1970s. Reaction to the 'social area analysis/factorial ecology/spatial behaviour' orientation (paralleled in other areas of human geography) has led to critiques of that mode of study and suggestions for new directions. Two such major reactions, leading in very different directions, can be identified.

From Opaque Description to Individual Actor

The first of these new directions was heralded by Buttimer (1969) with her reference to existentialists and the need to penetrate the 'subjective dimensions' (p. 417) of social space. During this period there was a considerable reaction to the dominance of a positivist philosophy in human geography (reviewed in Johnston, 1983), involving a desire to counter the 'dehumanising' nature of that orientation

> ... the method deployed by the sciences to understand man began to impinge on the value of being human. In a word, man was "naturalized" ... What the modern sciences demanded by the removal of all human elements (save, of course, rationality itself) was simply an agreement that man himself must be seen as a mere product of an environment (whether physical or social), a being understandable only in terms of phenomenal relations (Ley and Samuels, 1978, 7).

The subjective relationships that are the basis of links among individuals, places and environments were being ignored, and instead a pallid model of human beings was central to the research enterprise.

Within the field of social geography - as defined by Watson, Buttimer and Murdie - work (now generally termed humanistic geography) in the lacuna was initiated by David Ley. According to his arguments, the sort of geography reviewed above is an outsider's geography. But there is another (Ley, 1977a, 12)

> not the geography that we learn but the geography we come home to ... the world of everyday experience that is the ground for decision-making and action in all realms of behaviour.

His study of the insider's world is set in the tradition of J.K. Wright ('the geographical ideas, both true and false, of all manners of people'; 1947, 12) and D. Lowenthal ('We are all artists and landscape architects'; 1961, 260), but differences in its clearer links to philosophy. Ley draws on the phenomenology of Alfred Schutz and the concept of the 'taken-for-granted world', the created 'mental map' of the individual which structures behaviour that is habitual. The task of the social geographers is to understand this world, how it is created by the individual and how behaviour is organised within it (Ley, 1977b).

Ley's major empirical contribution to this mode of social geography was his doctoral dissertation - The Black Inner City as Frontier Outpost (Ley, 1973) - in which participant observation was the dominant methodology. The goal was to demonstrate that neither the 'objective' view of an area of inner Philadelphia, as derived from census and similar data, nor the 'external' views of white society (including that of 'white social science') were sufficient to understand the social geography of that small part of one city. 'Reality', as perceived locally, could only be appreciated through establishment of verstehen with the residents. Ley lived in the area and with its people, and from that experience developed a model of the black inner city as a frontier outpost, a territorial solution to the problems of living in an uncertain environment.

Ley's project (see Ley, 1981) is designed to present social geography as the study of places as meanings to people, not as mere mechanistic abstractions. Others have followed his lead, and have developed ideas relating to the nature of social life in its environmental context, as experienced by the individuals concerned (see, for example, some of the essays in Ley and Samuels, 1978, and Buttimer and Seamon, 1980): Duncan (1981), for example, has studied the role of the dwelling in social geography and the development of a personal 'sense of place' and the creation of residential landscapes to reflect the self-perceptions to groups within society (Duncan, 1973). Such studies illustrate the general directions being undertaken within the so-called humanistic perspective, though it is difficult to accept Jackson and Smith's (1984, 44) claim that they 'collectively refute the accusation that contemporary geographic humanism has produced an excess of exhortation and a dearth of applications'. In any case, at the same time as establishing a critique of positivist approaches, humanistic social geographers have been countering criticisms from an alternative perspective. They advance the study of places in which people live and to which they relate, but are attacked for failing to situate people within the economic and ideological frameworks that structure their interactions.

From Individual Actor to Structured Agent

The early 1970s saw the introduction to American geography of a strong Marxist realist component, focused around the publication of the journal Antipode at the Clark University Graduate School and the writings of an immigrant British scholar, David Harvey (1973). This development has been chronicled by Peet (1977), who saw it as a reaction to both national/international issues of the late 1970s and the failure of orthodox geography to grapple with the causes of major spatial and environmental issues.

Central to the Marxist argument is the view that individual actors operate within a clearly-articulated framework, the dynamics of the capitalist economic system. In the context of the traditional material studies by social geographers, this was made clearly by Harvey's (1974, 1975a, 1975b) analyses of the forces leading to residential differentiation within capitalist cities and the linked work by Walker (1978, 1981) on the processes of suburbanisation. In these works, the related processes of congregation and segregation were presented not simply as the outcome of choices, nor the manipulation of housing markets by institutional agents (e.g. Palm 1976), but as the consequences of economic processes deeply embedded in the American social formation. In a capitalist society, social differences are built upon economic variations; the conflict over where to live is closely linked to the contest for wealth, power and status within society, and the treatment of land and housing as commodities with exchange as well as use values ensures that this socio-economic conflict is also a spatial conflict. Further, because the state is fully implicated in the capitalist system, that spatial conflict is an an element of the political conflict that is central to capitalist society (Cox and Johnston, 1982).

The logic of this approach is that social geography cannot be separated analytically from economic and political geography. Social phenomena can provide the foci for empirical analyses but they cannot be understood without reference to the other spheres. For this reason, many of the major theoretical advances and essays - as in the books edited by Cox (1978) and Dear and Scott (1981) - have no specific social geographical focus; they are concerned with the totality of economic, social and political relations, whether concerned with the built environment in general (Harvey, 1978) or the living place in particular (Cox, 1981).

Whereas the humanistic approach focuses on the individual actors as creators of their own life-worlds, therefore, the so-called 'radical' approach focuses on the structure of society within which those actors operate. To some critics of the latter, this is even more dehumanising than the positivist approach: structural Marxism, it is claimed, presents the individual actor as little more than a 'bearer of the structure';

> the intersection of human geography with structural marxism has led to a passive model of man that is conservative and results in an obfuscation of the processes by which human beings can and do change the world ... the form of the explanations is both tautological and empirically untestable. The result is a mystification in explanation of how real processes operate (Duncan and Ley, 1982, 54-55).

Structural marxism is but one version of marxism, however - and is not adhered to, as described by Duncan and Ley, by one of the geographers they criticise - David Harvey (Johnston, 1985).

Very little Marxist-inspired work is as deterministic as suggested by such critiques. The more sophisticated analyses explore the interrelationships between structure and agency. Structures are created by humans (capitalism is a human invention), and they then act as 'containers' within which individuals are socialised and act. The processes of socialisation and action are activities undertaken by 'knowing' agents, who are free to interpret the demands of the capitalist system, within their acquired capabilities, and to act accordingly. This they do: the structure both constrains and enables action, and the action reproduces the structure. Thus society is continually being re-created, not as a series of carbon copies but as a sequence of interpretations.

These interpretations are structurally contained, and that containment, as Pred (1981) has argued, is spatial as well as social. Taking his stimulus from the time geography of Hagerstrand and the structuration theory of Giddens, Pred links individual to society by arguing that

> each and every person in his private life is affected by the public world, or society, around him. Yet each and every person is in some way a part of the public world, and thereby affects the private life content, or external actions and internal experiences of others. In order to gain insight into individual behaviour, it is necessary to hold insights as to the workings of society. In order to gain insights into the working of society, it is necessary to possess some insights into individual behaviour (pp. 231-232).

Such an approach is neither deterministic - in the sense applied by Duncan and Ley (1982) - nor idealistic/voluntaristic. People make, and are made by, societies.

The interactions between people and societies occur in places; as Pred's developments of time geography show, people are influenced in their contacts by where they are and the constraints on their movement. Socialisation is therefore spatially segmented, and can lead to spatial variations in interpretations of the economic imperatives of society, imperatives which are intergenerationally transmitted and become the bases of separate social formations - regional social and cultural geographies. This was argued by Peet (1975) in his analysis of poverty within the U.S.A. which identified spatial variations in poverty resulting from spatial variations in investment in social resources. It was extended in Soja's (1980) treatment of the socio-spatial dialectic, in which he argued that the production of space - i.e. various geographies - is itself a fundamental element of a capitalist system.

Empirical work in this framework has involved both the traditional foci of urban social geographers (as illustrated above with reference to Harvey's work on residential congregation and segregation) and new areas of study - such as that of gender (Zelinksy et al. 1982). These studies link social phenomena to the economic structure of society, and also to the political: the creation of social space involves political processes, for exammple, as Cox has shown in work on metropolitan local government fragmentation (Cox and Nartowicz,

1980) and conflict over neighbhourhood issues (Cox and McCarthy, 1982). Most attention has been paid to developing a theory of the economic organisation of capitalist society, however (notably in Harvey, 1982) and no attempt has been made to build a fully-fledged social geography within this perspective. (It could well be argued that a separate social geography is impossible, of course, but this should not preclude work which emphasised the social elements of the economic-political-social system.)

In Conclusion

Social geography has only recently developed within a discipline that focuses much more on artifacts than on people; indeed, one could construct a reasonable argument that recent American geography lacks any substantial social component. (This latter viewpoint is bolstered by the most recent review of post-1945 North American geography - Mikesell (1984) - which makes no explicit reference to social geography.) Only two attempts were made to define such a sub-discipline (by Wreford Watson, and by Buttimer) and neither had any substantial impact, although some of what Wreford Watson proposed was taken up in the technically sophisticated but theoretically naive work on urban residential patterns in the late 1960s/early 1970s. The intellectual failure of that enterprise, with its positivist emphasis on generalisations concerning spatial arrangements, was illustrated in the mid-1970s by the Association of American Geographers' Comparative Metropolitan Analysis Project. An atlas was produced (Abler, 1976) as part of this, but the greatest impact was the volume on urban social policy (Adams, 1976a) and the series of twenty 'vignettes' of selected urban areas (Adams, 1976b); each of the latter focused on the individuality of the place being described, and made little or no reference to general theories of socio-spatial organisation. To the extent that they were social geographies, the vignettes represented a return to the pre-1960s exceptionalism.

In 1983, however, the first explicit textbook on the subject appeared - David Ley's (1983) A Social Geography of the City. He criticises earlier approaches to urban geography for ignoring 'the experience, the social processes, and the broader contexts enveloping urbanisation', and seeks to create 'a distinctive social geography of the city' (p. vi) which integrates the previous emphasis on spatial form with work on everyday life, culture and values, social groups (informal institutional), and urban politics: his central focus is 'the quality of life experienced by urban residents'. Ley's philosophy of geography is strongly humanistic, and he advances a 'truly social geography' (p. 10) as one based in social phenomenology and pragmatism 'examining urban experience as a product of social and subcultural processes'. Thus

> we shall see that both the spatial form and the geographical experience of the city are the product of not only choice but also constraint, the outcome of private ambitions, mutual objectives, interest-group power, and historical possibility (p. 13).

To those of marxist and other realist persuasions, the problem with this approach is that it pays insufficient attention to how choices and constraints are produced; it portrays human agency in action without analysing the structure within which the action is enabled.

North American geography has been dominated in recent decades by a concern with what is where, with the artifacts produced by society rather than with society itself. Social geography, the study of 'how social life is constituted geographically through the spatial structure of social relations' (Jackson and Smith, 1984, vii), was thus of little interest. The availability of census data and the technology to manipulate allowed the brief appearance of an urban social geography focused on the residential congregation and segregation of people occupying different social positions but this, and the associated behavioural work, failed to explore either the experience of urban social life or the causes of the segregated city. More recently, both of these lacunae have been tackled, and the expectation is that in the near future social geography - emphasising the fundamental role of structured social relations in the creation and re-creation of environments - will come to occupy the central stage.

References

Abler, R.F. 1976 A Comparative Atlas of America's Great Cities. University of Minnesota Press, Minneapolis

North America

Adams, J.S.	1972	The geography of riots and civil disorders in the 1960s, Economic Geography, 45, 302-232
Adams, J.S. editor	1976a	Urban Policymaking and Metropolitan Dynamics. Ballinger, Cambridge, Mass
Adams, J.S. editor	1976b	Contemporary Metropolitan America: Twenty Geographical Vignettes. Ballinger, Cambridge, Mass
Bach, W.	1971	Geography of Air Pollution. McGraw Hill, New York
Barrows, H.H.	1923	Geography as human ecology. Annals of the Association of American Geographers, 13, 1-14
Berry, B.J.L.	1964a	Approaches to regional analysis: a synthesis. Annals of the Association of American Geographers, 54, 2-11
Berry, B.J.L.	1964b	Cities as systems within systems of cities. Papers and Proceedings, Regional Science Association, 13, 147-163
Berry, B.J.L.	1965	Research frontiers in urban geography. In P.M. Hauser and L.F. Schnore, editors. The Study of Urbanization. John Wiley, New York, 403-430
Berry, B.J.L.	1971	Comparative factorial ecology. Economic Geography, 47, 209-367
Brown, L.A.	1968	Diffusion Processes and Location. Regional Science Research Institute, Philadelphia
Brown, L.A.	1982	Innovation Diffusion: A New Perspective, Methuen, London
Brown, L.A. & E.G. Moore	1969	The intra-urban migration process. Geografiska Annaler, 52b,

		1-13
Buttimer, A.	1968	Social geography. In D.L. Sills, editor, _International Encyclopedia of the Social Sciences_, Vol. VI, Macmillan, New York, 134-145
Buttimer, A.	1969	Social space in interdisciplinary perspective. _Geographical Review_, 59, 417-426
Buttimer, A.	1971	_Society and Milieu in the French Geographical Tradition_. Rand McNally, Chicago
Buttimer, A. & D. Seamon	1980	_The Human Experience of Space and Place_. Croom Helm, London
Cox, K.R.	1969	The voting decision in a spatial context. In C. Board et al. editors _Progress in Geography_, 81-117
Cox, K.R.	1973	_Conflict, Power and Politics in the City_. McGraw Hill, New York
Cox, K.R. editor	1978	_Urbanization and Conflict in Market Societies_. Methuen, London
Cox, K.R.	1980	_Location and Public Problems_. Basil Blackwell, Oxford
Cox, K.R.	1981	Capitalism and conflict around the communal living space. In M.J. Dear & A.J. Scott, editors, _Urbanization and Urban Planning in Capitalist Society_, Methuen, London, 431-456
Cox, K.R. & R.G. Golledge editors	1969	_Behavioural Problems in Geography_. Northwestern University Department of Geography, Studies in Geography 17, Evanston
Cox, K.R. & R.J. Johnston	1982	_Conflict, Politics and the Urban Scene_.

editors		Longman, London
Cox, K.R. & J.J. McCarthy	1982	Neighbourhood activism as the politics of turf. In K.R. Cox & R.J. Johnston, editors Conflict, Politics and the Urban Scene, Longman, London, 196-219
Cox, K.R. & F. Nartowicz	1980	Jurisdictional fragmen-fragmentation in the American metropolis: alternative perspectives. International Journal of Urban and Regional Research, 4, 196-211.
Davies, W.K.D.	1984	Factorial Ecology. Gower, Aldershot
Dear, M.J. & A.J. Scott editors	1981	Urbanisation and Urban Planning in Capitalist Society. Methuen, London
Duncan, J.S.	1973	Landscape taste as a symbol of group identity. Geographical Review, 63, 334-355
Duncan, J.S.	1980	The superorganic in American cultural geography. Annals of the Association of American Geographers, 70, 181-198
Duncan, J.S. editor	1981	Housing and Identity. Croom Helm, London
Duncan, J.S. & D. Ley	1982	Structural marxism and human geography: a critical assessment. Annals of the Association of American Geographers, 72, 30-59
Duncan, O.D. & H.W. Pfautz	1960	Translation of M. Halbwachs Morphologie Sociale. The Free Press, Glencoe, Ill
Duncan, O.D. & B. Duncan	1955	Occupational stratification and residential distribution. American Journal of Sociology, 50, 493-503
Duncan, O.D. &	1959	Cultural, behavioural

North America

L.F. Schnore		and ecological perspectives in the study of social organisation. American Journal of Sociology, 65, 132-146
Entrikin, J.N.	1980	Robert Park's human ecology and human geography. Annals of the Association of American Geographers, 70, 43-58
Eyles, J.D. & D.M. Smith	1978	Social geography. American Behavioural Scientist, 22, 41-58
Goheen, P.G.	1970	Victorian Toronto, 1850-1900. University of Chicago, Department of Geography, Research Paper 127, Chicago
Golledge, R.G., L.A. Brown, & F. Williamson	1972	Behavioural approaches in Geography: an overview. The Australian Geographer, 12, 59-79
Harries, K.D.	1974	The Geography of Crime and Justice. McGraw Hill, New York
Harris, G.D. & E.L. Ullman	1945	The nature of cities. The Annals, 242, 7-17
Hartshorne, R.	1939	The Nature of Geography. Association of American Geographers, Lancaster, Pa
Hartshorne, R.	1959	Perspective on the nature of Geography. Rand McNally, Chicago
Harvey, D.	1973	Social Justice and the City. Edward Arnold, London
Harvey, D.	1974	Class-monopoly rent, finance capital and the urban revolution. Regional Studies, 8, 239- 255
Harvey, D.	1975a	Class structure in a capitalist society and the theory of residential differentiation. In R. Peel et al. editors, Processes in Physical and Human

		Geography: Bristol Essays, Heinemann, London, 354-469
Harvey, D.	1975b	The political economy of urbanisation in advanced capitalist societies. In G. Gappert & H.M. Rose, editors, The Social Economy of Cities, Sage Publications, Beverly Hill, 119-163
Harvey, D.	1978	Labor, capital and class struggle around the built environment in advanced capitalist societies. In K.R. Cox, editor, Urbanization and Conflict in Market Societies, Methuen, London, 9-37
Harvey, D.	1982	The Limits to Capital. Basil Blackwell, Oxford
Hawley, A.H.	1950	Human Ecology. The Ronald Press, New York
Jackson, P. & S.J. Smith	1981	Introduction, in P. Jackson & S.J. Smith, editors, Social Interaction and Ethnic Segregation, Academic Press, London, 1-18
Jackson, P. & S.J. Smith	1984	Exploring Social Geography. George Allen and Unwin, London
James, P.E. & C.F. Jones editors	1954	American Geography: Inventory and Prospect. Syracuse University Press, Syracuse
James, P.E. & G.J. Martin	1981	All Possible Worlds. John Wiley, New York
Johnston, R.J.	1983	Geography and Geographers. Edward Arnold, London.
Johnston, R.J.	1985	David Harvey's geography ... Antipode
Kuhn, T.S.	1962	The Structure of Scientific Revolution. University of Chicago Press, Chicago
Ley, D.	1973	The Black Inner City as

		Frontier Outpost Association of American Geographers, Washington DC
Ley, D.	1977a	The personality of a geographical fact. *The Professional Geographer*, 29, 8-13
Ley, D.	1977b	Social geography and the taken-for-granted world. *Transactions, Institute of British Geographers*, NS2, 498-512
Ley, D.	1981	Behavioural geography and the philosophies of meaning. In K.R. Cox & R.G. Golledge, editors, *Behavioural Problems in Geography Revisited*, Methuen, London, 209-230
Ley, D.	1983	*A Social Geography of the City*. Harper and Row, New York
Ley, D. & M.S. Samuels, editors	1978	*Humanistic Geography*. Croom Helm, London
Lowenthal, D.	1961	Geography, experience and imagination: towards a geographical epistemology. *Annals of the Association of American Geographers*, 51, 241-260
Mayer, H.M.	1954	Urban geography. In P.E. James & C.F. Jones, editors, *American Geography: Inventory and Prospect*, Syracuse University Press, Syracuse, 142-166
Mayer, H.M. & C.F. Kohn, editors	1959	*Readings in Urban Geography*. University of Chicago Press, Chicago
Mikesell, M.W.	1984	North America. In R.J. Johnston & P. Claval, editors, *Geography since the Second World*

		War: An International Survey, Croom Helm, Beckenham, 185-213
Morrill, R.L.	1965	The Negro ghetto. Geographical Review, 55, 339-361
Morrill, R.L.	1984	Reflections of the quantitative revolutions early years: Washington 1955-1965. In M. Billinge et al., editors, Reflections of a Revolution, Macmillan, London, 57-72
Morrill, R.L. & E.H. Wohlenberg	1971	The Geography of Poverty in the United States. McGraw Hill, New York
Murdie, R.A.	1969	Factorial Ecology of Metropolitan Toronto 1951-1961. University of Chicago Department of Geography, Research Paper 116
Murdie, R.A.	1976	Spatial form in the residential mosaic. In D.T. Herbert & R.J. Johnston, editors, Social Areas in Cities: Spatial Processes and Forms. John Wiley, Chichester, 237-272
Murphy, R.E.	1966	The American City, McGraw Hill, New York
Palm, R.I.	1973	Factorial ecology and the community of outlook. Annals of the Associaton of American Geographers, 63, 341-346
Palm, R.I.	1976	The role of real estate agents as information mediators in two American cities. Geographiska Annaler, 58B, 28-41
Peet, J.R.	1975	Inequality and poverty: a marxist-geographic theory. Annals of the Association of American

		Geographers, 65, 564-571
Peet, J.R.	1977	The development of radical geography in the United States. Progress in Human Geography, 1, 240-263
Pooler, J.A.	1977	The origins of the spatial tradition in geography: an interpretation. Ontario Geography, 11, 56-83
Pred, A.R.	1981	Of paths and projects. In K.R. Cox & R.G. Colledge, editors, Behavioural Problems in Geography Revisited, Methuen, London, 231-255
Porter, P.W.	1978	Geography as human ecology. American Behavioural Scientist, 22, 15-40
Pyle, G.F.	1971	Heart Disease, Cancer and Stroke in Chicago. University of Chicago, Department of Geography, Research paper 134
Redford, J.P.	1981	The social geography of the nineteenth century U.S. city. In D.T. Herbert & R.J. Johnston, editors, Geography and the Urban Environment, Volume 4, John Wiley, Chichester, 257-293
Rees, P.H.	1970	Concepts of social space: towards an urban social geography. In B.J.L. Berry & F.R. Horton, Geographic Perspectives on Urban Systems, Prentice-Hall, Englewood cliffs, NJ, 306-394
Robson, B.T.	1969	Urban Analysis. The University Press, Cambridge
Rose, H.M.	1970	The development of an

		urban subsystem: the case of the Negro ghetto. Annals of the Association of American Geographers, 60, 1-17
Rose, H.M.	1971	The Black Ghetto: A Spatial Behavioural Perspective. McGraw Hill, New York
Rushton, G.	1969	The analysis of spatial behaviour by revealed space preferences. Annals of the Association of American Geographers, 59, 391-400
Schnore, L.F.	1961	Geography and human ecology. Economic Geography, 37, 207-217
Schnore, L.F.	1965	On the spatial structure of cities in the two Americas. In P.M. Hauser & L.F. Schnore, editors, The Study of Urbanization, John Wiley, New York, 347-398
Shannon, G.W. & G.E.A. Dever	1974	Health Care Delivery: Spatial Perspectives. McGraw Hill, New York
Smith, D.M.	1973	The Geography of Social Well-Being in the United States. McGraw Hill, New York
Soja, E.W.	1980	The socio-spatial dialectic. Annals of the Association of American Geographers, 70, 207-225
Vise, P. de	1973	Misused and Misplaced Hospitals and Doctors. Resource Paper 23, Commission on College Geography, Association of American Geographers, Washington DC.
Walker, R.A.	1978	The transformation of urban structure in the nineteenth century and beginnings of suburbanisation. In K.R.

		Cox, editor, <u>Urbanisation and Conflict in Market Societies</u>, methuen, London, 165-212
Walker, R.A.	1981	A theory of suburbanisation. In M.J. Dear & A.J. Scott, editors, <u>Urbanisation and Urban Planning in Capitalist Society</u>, Methuen, London, 383-430
Ward, D.	1971	<u>Cities and Immigrants</u>. Oxford University Press, New York
Watson, J. Wreford	1953	The sociological aspects of geography. In G. Taylor, editor, <u>Geography in the Twentieth Century</u>, Methuen, London, 463-499
Whittlesey, D.	1954	The regional concept and the regional method. In P.E. James & C.F. Jones, editors, <u>American Geography: Inventory and Prospect</u>. Syracuse University Press, Syracuse, 18-69
Wright, J.K.	1947	Terrae Incognitae: the place of imagination in geography. <u>Geographical Review</u>, 37, 1-15
Zelinsky, W.	1966	<u>A Prologue to Population Geography</u>. Prentice-Hall, Englewood Cliffs NJ
Zelinsky, W.	1973	<u>The Cultural Geography of the United States</u>. Prentice-Hall, Englewood Cliffs, NJ
Zelinsky, W. et al	1982	Women and geography. <u>Progress in Human Geography</u>, 6, 317-366

Chapter Three

BRITAIN

John Eyles

Introduction

While this chapter will not equate social geography with the whole of human geography as was suggested in the 1940s (Fitzgerald, 1946), it will range far and wide in its attempt to describe and understand the nature of social geography. It will try and draw parallels between the research carried out in early and contemporary social geography. In this, it will be recognised that academic ideas, like those in everyday life, come from somewhere and have a context, but that disciplines develop in an evolutionary fashion not in stages so that there are 'relict' topics and approaches and continuations of the old alongside the new (Eyles, 1984). It will consider non-British material in that its antecedents are in the main French and American and the non-geographical in that its dominant ideas to-day seem derived from social and political analysis. To be churlish, it is possible to define away any original contributions for British social geography. But that would miss the point, because it has modified and applied foreign ideas to its own empirical practice. Further, the synthesising tradition of British geography has meant that the numerically and ideationally dominant American endeavours have been subjected to an adjectival imperialism in which social geography is seen as a predominantly British phenomenon.

This conclusion is possible because of the lack of American interest in the field (see Johnston, loc. cit.) and despite the French origins of the term (Dunbar, 1977). The term originates from around 1884 when Reclus used it to describe the study of man and nature in intricate relation. His work, while emphasising the unequal distribution of resources and wealth, becomes a study without geographical context in its consideration of modern times (Dunbar, 1981). It may be said that the past 100 years of British social geographical endeavour has been an eclectic search for ways of ensuring a lack of geographical content. Such an assertion is unfair, but we may ask what has been the purpose of its eclecticism.

It is, however, incorrect to suggest that the influence of Reclus' ideas has been particularly strong in Britain or France. The same can be said for what seems to be the first explicit use of the term in British sources by the American, Hoke (1907). He decribes social geography as being the examination of

> the distribution in space of social phenomena. Its working programme may be stated as the 'description of the sequence and relative significance' of those factors, the resultant of whose influences is the localisation in space of the series of social phenomena chosen for investigation .... The subject deals, therefore, with the facts and products of human association as represented by group characteristics, industries, technology, customs, beliefs and related phenomena; and estimates the significance of the various forces which have influenced their distribution (Hoke, 1907, 64-5).

It was, however, the ideas of other scholars, particularly Le Play and Vidal de la Blache, that had a greater impact on the development of British social geography.

French Ideas and Inter-war Social Geography

The tensions, real and imagined, between the two 'explanatory schools' of geography, determinism and possibilism, may be discerned in the sub-discipline through the incorporation of the ideas of Le Play and Vidal. 'Tensions' perhaps implies a far too conscious activity because few seemed concerned enough to state whether they were one or the other.

The ideas of Vidal are, of course, of great significance for all geographers. He viewed geography as a natural and a human science, a scientific subject studying all the phenomena of the earth's surface which were seen as existing as a unity

(Vidal de la Blache, 1913; Harrison Church, 1957). The works of man rather than the influence of environment were, however, the starting-point. Nature was seen as an adviser, a provider of possibilities with man to judge their use (Febvre, 1925) and a determinant of only those limits beyond which life is impossible (Brunhes, 1920).

Such possibilist statements indicate that while human activity may be circumscribed by the environment, it is not 'fatally determined' as Brunhes put it. There is, however, no clear-cut division between people and environment for the simple reason that commonsense tells us that such a division would be meaningless. Thus, the language of the possibilists becomes similar to that of the determinists with Brunhes writing about 'the geographical imprisonment of our sedentary life' and Vidal of 'the sovereign influence of the environment' (see Tatham, 1957). Further, McKenzie (1924) in his explanation of the ecological approach - a set of ideas not greatly influential in Britain until the post-1945 period - could approvingly cite Brunhes' work for identifying the physical necessities of house, road and water as being three essentials for the inception of the human community. Such examples point to the problem of eclecticism in that the context of the borrowed idea may not be fully appreciated or realised (see Agnew and Duncan, 1981). They also demonstrate how theoretical rigour tends to get smudged as a theory comes up against the real world. Its conceptual accentuations, directions and orientations may not hold. Clarity in theory becomes muddied by empirical investigation, as we may also note in the context of the Chicago School, where the 'rediscovery' of its pragmatic and humanistic bases may well point to a disjuncture between its theory and practices rather than the conceptual missing link between the School's case studies and modern social geographical theory.

In the practice of inter-war social geography, it is perhaps the Vidalian emphasis on regional description rather than the French School's epistemological base that was developed. Such development may be seen as being in accord with the empirical tradition of scientific enquiry in Britain. This tradition may also explain the significance of Le Play's ideas for if Vidal assisted in the establishment of the arena of investigation, Le Play provided a framework and methodology. Indeed as Watson (1957) points out, the advance of social geography probably owes more to Le Play than anyone else. Le

Play argued that _lieu_, _travail_ and _famille_ were necessary concepts for understanding the nature of society. He argued that ecology was the most important factor for this understanding (see Sorokin, 1964), and some of those inspired by his work, whom Roxby (1930) called 'geographical sociologists', stressed the control of human activities by natural conditions. Le Play focussed his own attention on the family and saw its struggle for survival, security and happiness as the key to understanding social change. But _lieu_, _travail_ and _famille_ were seen as being in intricate relation with one another.

The general themes of Le Play's work were taken up in Britain by Geddes (Freeman, 1965; Robson, 1981), particularly the theme that environment and mode of living in an area are reflected in common social attributes and in the mores and fabric of the family. Geddes broadened the focus of attention from family to folk (place-work-folk) and from the method of the family budget to the social survey. Although he broadened the debate, his impact, and through him that of Le Play, has been anticlimactic. While he demonstrates the practicality and pragmatism of British science and social geography by regarding the controversy of determinism and possibilism as futile -

> the human group is influenced by its environment but learns to dominate it; by improving its tools and weapons it creates a store of free energy that is available for art and thought and play; and it criticises its laws and customs and modifies its social heritage by deliberate selection. Hence human society comprehends polity, culture and art as well as work, folk and place (quoted in Freeman, 1965, 87-8) -

his organic theorising, social analysis and political and mystical thinking has been lost. In Robson's terms, the legacy is diluted and what were passed on were his survey method and regional interests.

'Regional survey' is best exemplifed in social geographical terms by the work of Fleure. But to see Fleure simply as a regionl surveyor is to lose his significance for the subject's development in Britain. A zoologist by training, Fleure advocated an evolutionary approach to the condition of man and culture in the regions of the world. While such an approach seems Darwinian, his emphasis on holism (see Fleure, 1929) appears full of Vidalian possibilities. It seems that Geddes was right about the futility of the debate.

Fleure argued that people should not be divorced from the study of their environments but the latter were seen in terms of cultural appraisal. Such an interpretation resulted in a strong link being forged between social geography and anthropology, as seen in Fleure's own work on racial characteristics (see Fleure and James, 1916). The link stemmed in part from the question raised by the possibilist reaction to environmental determinism, namely the problems of human adjustment to particular environmental conditions through different modes of living. Fleure's successor to the chair at Aberystwyth provided a succinct summary;

> Between the physical environment and human activity there is always a middle term, a collection of specific objectives and values, a body of knowledge and belief: in other words, a cultural pattern. That the culture is ... adaptable and modifiable in relation to physical conditions, must not be allowed to obscure the fact that adaptation proceeds by discoveries and intentions, which are in no sense inevitable. Equally important are the restrictions placed by social patterns and religious concepts on the utilisation of resources or on adaptations to physical conditions (Forde, 1934, 463).

This fifty year old statement has a particularly modern ring; especially if the terms change (adaptation), ideology (social patterns and religious concepts), everyday life (culture), and struggle, materiality and consciousness (discoveries and inventions) are utilised. But adaptation is examined in terms of cultural differentiation, or different cultural responses as manifested in a way of life to environing conditions. There could not be more implicit support for the Vidalian view.

Nor indeed could there be more explicit advocacy than in Roxby's (1930) view that social geography is primarily concerned with the regional distribution and interrelation of different forms of social organisation developing from particular modes of life. It is a holistic science (see Kinvig, 1953).

Roxby's (1925) own work on the relationships between population growth and density and standard of living in China may be cited as an example. So too may Fleure's (1919) seminal study of regional classification based on man's efforts to pursue his 'social well-being'. He identified regions of hunger, debilitation, increment, effort, difficulty and wandering as well as industrialised regions. The keynote of the paper is 'human experience'. It emphasises "the importance of human will and man's power to modify his surroundings" (Fleure, 1919, 94). It stressed the dynamism of the relationship between people and environment and, like Roxby, Fleure points to the relations between geography and the other sciences, specifically anthropology and history.

The anthropological relationship came at a time of massive functionalist field surveys of small-scale societies (see Pahl, 1965a; Lewis, 1976). In such societies, the relations between peoples and environments are more immediate and visible then they tend to be in urban-industrial societies. The 'personality' of a place and a people are thus more readily discerned in communities where the mode of living appears as a direct response to environmental possibilities, as, in particular, in agriculture. The significance of this anthropological perspective, reviewed and summarised by Forde (1934), in emphasising the totality of life-in-a-place may be seen in the studies of rural communities in Britain and Ireland. Such communities again possessed a recognisable and identifiable holism which focused on self-sufficiency in social, economic and affective matters.

The department at Aberystwyth instigated a 'programme' of investigating the "enduring relationship between society and the physical environment" (Williams, 1963, xx). The most important study was perhaps the first by Rees (1950) in the parish of Llanfihangel where life was firmly based on the land and agriculture and where depopulation was impoverishing the culture. The presence of kindred and the need for economic cooperation dominated the locality. Work, place and folk were complexly interwoven.

> Each farm is the centre of a circle of cooperators which differs slightly from those of its neighbours while overlapping with them, and in this way the whole countryside is covered by a continuous

network of reciprocities (Rees, 1950, 94).

The isolation and remoteness of Llanfihangel was breaking down even in the 1940s. The cultural adjustments made by various Welsh communities were in fact emphasised by other researchers (see Jenkins, et al, 1960). Even so, there remains an emphasis on the traditional as opposed to the emerging aspects of social processes (see Jones, 1980). Thus, in Glan-Llyn, it was concluded that much of the peasant way of life was maintained and in Aberdaron that ancient loyalties survived serving as a token of the unbroken unity of the area from the earliest times to the present day (Jenkins, et al, 1960). It may well be that the dominance of 'environment' meant that continuity rather than change would be emphasised. Indeed, in his two studies of English villages, Williams (1960; 1963), trained by Rees, noted that in Gosforth the relationships of the elementary family were still enmeshed in a network of ties based on proximity and consanguinity, and that in Ashworthy the continuity of the family farm manifested the relationships between society and environment over time. In both instances, the changes resulting from different attitudes, techniques and organisations from the 'outside' are commented on as phenomena altering the social and ecological nature of these rural communities. These phenomena were to dominate the sociological studies of the rural world (see, for example, Newby, 1977; Strathern, 1982).

Indeed these changes which saw Britain become a more mobile yet highly integrated and obviously urbanised society saw the demise of the anthropologically-influenced, holistic analyses of man-land relationships. These analyses were eventually replaced by sociological studies emphasising structure and institutions. And a geographer, turned sociologist, wrote their cruellest epitaph:

> In a sense the anthropo- or socio-geographic study of a modern rural community was not a great advance on studies of the traditional peasant communities in the 'Man and his conquest of Nature' school. Certainly kinship analysis may have been substituted for household analysis as a meaningful clue to the 'adjustment' of the social group to the land, but in an <u>urban</u> environment the relationship with the physical environment may be completely

without significance (Pahl, 1965a, 83-4).

Such an epitaph must be challenged. Its conscious or unconscious acceptance, related to the emergence of other investigative interests, resulted in a downgrading, if not rejection, of environment in social geography beyond the level of pattern identification (see Eyles, 1985) and helped to instigate the eclecticism in the explorations and explanations of the social geographer-as-sociologist, with the subject becoming sociology or political economy. And such acceptance has meant the diminution of the ideas of Fleure and his students. Society-environment relations were much to the fore in their studies as may be seen in the intriguingly entitled collection of essays presented to Fleure, namely studies in regional consciousness and environment (Peate, 1930) and particularly in the work of Evans (1942; 1951). It is exceedingly difficult to discover where any writer stood on the issue of, say determinism and possibilism or on the role of 'society'. Such questions and labels were unnecessary. Thus, it is the acknowledgements that we learn that Evans' (1951) aim was to trace in Mourne Country, a small and well-loved region, the physical components and human responses which make up the Irish countryside. To this end, we are treated to descriptions of its landscape, early history, climate and vegetation, and human occupance and responses in such chapters, 'giants' graves', 'raths and saints', 'Normans and planters', 'wrack harvest', 'pedlars and smugglers', 'house and hearth', and 'the elder faiths'. Further, it is in the preface of Jones (1960; xiii), a student of Forde and in a book which links this tradition with the interest in the city, that we read that "the landscape is as much the outcome of social values as of physical background and history".

This 'loss' of these ideas has not meant interest in society-environment relations has been completely terminated. Anthropologists have examined how broad social processes affect rural and marginal communities (see Cohen, 1982) and how such relations are played out in the city (see Hannerz, 1980; Susser, 1982, although these are American examples). To anticipate, it is the potentiality of this anthropology for social geography which points to the continued relevance of inter-war studies and the idea of Fleure and Evans (and hence Vidal and Le Play) for the analyses of society-environment relations and of regional or locality consciousness.

American Ideas and the Scientific Investigation of the City

Although begun before the Second World War, the first text on urban geography in English appeared in 1946 (Taylor, 1946). Under German influence, urban studies concentrated on the historical and economic (morphological and functional) rather than the social, although Taylor's approach seems similar to that discussed above, seeing the city as an evolving organism. And while we should not too readily draw parallels between different branches of thought, such a view of the city was also held by Park and other human ecologists whose work at Chicago was to be so influential in shaping the analyses of urban social geographers. But their interests in the social nature of urban areas in the 1920s was not matched in Britain until the 1950s and 1960s. Lack of interest in the city in so early an urbanised society as Britain may be due to the anti-urbanist trends in thought and practice identified by Glass (1955). It may also be that social problems and differences were seen in industrial and economic terms. So Park's ideas were first introduced to British geographers by Dickinson (1957) and later elaborated by Pahl (1965a) and Robson (1969).

Their discussions on the nature of human ecology pointed to the Darwinian borrowings of Park, specifically that the starting point of ecological theory was the intimate relationships between organisms and between organism and the environment. Man was subjected to the general laws of the natural world, but was not so dependent on the physical environment as other animals. Custom and culture limited the competition in the biotic world (Park, 1936). The concepts of the human ecologists - competition, invasion, succession and dominance - have been used, often in modified form, to investigate neighbourhood change, residential mobility and segregation. Peach (1975; 1981) regards Park's correlation of social with spatial relations and distance as being a major argument for applying spatial physics notions to the study of segregation. Jackson and Smith (1981) have a different interpretation of Park's ideas emphasising his association with Chicago ethnographers and interactionists and his ideas on anthropology, morality and history. This interpretation, while doing little to rehabilitate Park's theorising, represents some of the Chicago School's practice, the rich tapestry of studies on urban structure, process and psychology, which again stress the society-environment relation.

The School's theorising is largely deterministic (see Timms, 1971). As Ley (1983) reminds us, Park's alternative title for human ecology was biological economics. The theoretical complexities of the cultural are avoided and community - natural area - is a human aggregation living in a fixed geographical locale and controlled by the subcultural processes of competition, cooperation, assimilation and conflict. It is perhaps unfortunate that Park's rather than Wirth's (1938) ideas have been so dominant in social geography. Wirth's discussions on the social nature of urbanism have perhaps only been successfully updated by Raban (1975), a journalist, and it may be that the criticisms of Wirth by sociologists of such stature as Gans (1962) deflected attention, although Wirth's deductive generalisations did not sit easily with the dominant social geographical endeavour of the 1960s, that of technicalist inductive analyses.

But Park's ideas, the least social of all theories, was implicitly adopted by social geographers in their use of ecological analysis. If the causation of action by environment is denied, a strong and unproblematic associaton is assumed. Ecological analysis became central and as Reissman (1964, 97-8) points out such study of the environment includes

> a census of the species and organisms it contains, a tracing of the lines of interdependence, and a comparison over time of changes that have occurred.

Its attraction lies in its simplicity in that purpose, motivation and interest can be ignored in favour of the derived patterns. Cities were therefore, described and analysed in terms of land uses and populaton attributes. Such analyses became more attractive with the post-war availability of census enumeration district data as studies of Middlesborough (Glass, 1948) and Oxford (Mogey, 1956) attest and the application of computer-based multivariate analyses of which Moser and Scott's (1961) classification of British towns is an early example of studies later dominated by the intra-urban scale. And can there be a better description of what factor ecologies do than Reissman's statement?

The availability of data and techniques of analysis led to a dearth of theoretical interest, although ecological notions remained important. It allowed regional description and areal differentiation (combined as pattern identification) to take centre-stage unencumbered by questions concerning the nature of man and his relationship to the environment. Multivariate, ecological analyses produced in the main regional geographies of the city with the intuitive bases of investigation, as examplified by Watson's (1957) study of Hamilton, Ontario, being replaced by the census derived variables of factorial ecology. (These analyses also prevented social area analysis, the macro-social criticism of the Chicago School which started with the relationship between the city and its encompassing society, being widely used in Britain: but see Herbert's (1967) study of Newcastle-under-Lyme.) But Watson's description of his own work is applicable to the whole field. He saw social geography as being concerned with the identification of different regions according to associations of social phenomena relating to the total environment. It emphasised the geographical expression of social differences rather than a geography of society, concentrating on the place of Le Play. If 'place-work-folk were all considered, the study would become sociology.

An unproblematic relation between ecological and social structure is assumed and the sophistication of the mode of analysis allowed that assumption to be ignored for some 20 years. The questioning of the relation did not usually come in the factorial ecologies themselves, which are surveyed by, among others, Rees (1970), Timms (1971) Johnston (1976), and Herbert andd Thomas (1982). Factorial ecologies were noted for the eclecticism of technique, the choice of which is important for determining the actual statistical relationship between variables (see Johnston, 1976). Statistical explanation often seemed to be elided into substantive explanation, although the latter kind of association was usually highly speculative. Differences were noted between the British and North American experience and were explained mainly in terms of data differences and the role of public housing in shaping urban structure. Factorial ecologies remained virtually untouched by the emerging urban social theory concerning social and spatial interactions in the socio-ecological system (Pahl, 1968) that would result in the development of the managerialist thesis. Indeed ecological analyses were more tellingly questioned by the earlier investigations of Belfast (Jones, 1960) and Sunderland (Robson, 1969).

Britain

Jones' seminal work on Belfast has already been mentioned in the context of the social geography of Fleure and Evans. It also utilised census data and applied ecological models, albeit those modified by Firey (1947). It could be said to stand at the crossroad for several strands of development, but its lasting signicance is its demonstration of the role of sentiments and symbols (i.e. social values) in shaping urban land-use patterns and social activities. Theories of urban structure, e.g. the ecological, which imply a mechanistic adjustment of society within the urban framework, do not hold. Historical analysis demonstrates the role that social values played in determining what is where and who lives where. Such notions are supported again in the Belfast context by Boal (1969).

If Jones' added historical analysis to confound ecological theory, Robson (1969) utilised survey analysis to enable him to conclude that residential location is affected by the age structure of the population, the importance of the local area to the individual's social world and the significance of 'milieu' to urban social structure. Milieu is taken to mean not only housing and environmental conditions but also relative location and psychological factors. This emphasis on the behavioural is interesting and Robson's (1969, 264) desire to throw light on the structure of the British city by fusing "the spatial interests of geography and the structural interests of sociology ... to explore some of the relationships between space and social structure" provides an early pointer to the subject's theoretical debate.

Values, milieu and behaviour are also found in Pahl's (1965b) work on the metropolitan fringe. Pahl, a student of Jones, looked at the ways in which social perceptions affected the interactions between different classes in specific locales. But the locality is now seen as a stage on which social processes are played out. Pahl's work becomes overtly sociological, emphasising interactions and institutions in an urbanised world - the social worlds of exurbia. While research on the metropolitan fringe continued (see Connell, 1974), the interactionist and institutional perspectives did not appear to contain a geographical dimension - until the late 1970s at least. Further, the sociological imperialism of the study of social relationships denied a significant role for the location of such relationships. The demise, or rather destruction, of community studies (see Stacey, 1969) had the consequence of divorcing geography and sociology at a meaningful empirical level - locality and neighbourhood. For ten years or so, social geographers were exposed to social theories which were only of apparent marginal relevance to their reality (see Pahl, 1967; Eyles, 1974), although these expositions did point to the underlying assumptions of social theories and models. But in the main, social geographers got on with their studies of residential differentiation or turned to behaviour.

More American Ideas and the Incorporation of Behavioural Variables

The incorporation of behavioural variables in social geography occurred at roughly the same time as the quantifiction of the field (see Eyles and Smith, 1978) and resulted from a dissatisfaction with optimising, omniscient economic man as a model of behaviour. The behavioural locational principles introduced were, however, still predicated on quasi-ecological and neo-classical economic reasoning. Individuals were seen as possessing individual attributes - age , sex, socio-economic status - but were still passive, unconscious actors. Individuals were seen as satisficing decision-makers with limited abilities to handle information.

Research focussed on environmental cognition. In many ways, this field, first developed in the US, seemed an exciting advance by trying to answer the questions of how we know the world and how we learn about that which is taken-for-granted. In reality, in both the US and the UK, behavioural concerns in social geography become the arena for further methodological eclecticism with major borrowings from social psychology. There have been some attempts to model the process of environmental cognition (Pocock and Hudson, 1978; Gold, 1980) but the utilisation of technique has tended to abstract the image or information from the context of individual lives and the environmental image from the total image structure. Thus there have been technically sophisticated studies of images of the retail environment (Downs, 1970; Hudson, 1974); of the use of personal construct theory (Harrison and Sarre, 1971); of information flows (Goddard, 1978) and of spatial search (Lloyd and Dicken, 1972; Hudson, 1976). There have also been studies that have emphasised the derivation and form of environmental images and preferences (e.g. Gould and White, 1968;

Herbert, 1975). Interesting findings have been forthcoming. Recent migrants have restricted urban images (Spencer and Weetman, 1981) while the wealthy have extensive ones (Walmsley and Lewis, 1984). But technique has been the driving force with Spencer (1973) in fact suggesting that the delimited image varies according to the technique adopted: paradoxically, unique outcomes to such eclecticism.

The parallels that have been drawn been this behavioural research and the insights of Kirk (1951, 1963) are therefore erroneous. At a very general level, there are similarites but Kirk used Gestalt psychology to differentiate the phenomenal environment - the natural and man-made world - from the behavioural - a psycho-physical field in which phenomenal facts are arranged into patterns and acquire values in cultural contexts. The environments are mutually interactive. Kirk's work should perhaps be seen as the end of an era, that which emphasised society and environment in intricate interrelation, rather than the start of a technicialist behavioural geography. That is not to say that Kirk's work is without contemporary relevance. It can form a focus of an emerging critical social geography.

The notions concerning environmentl cognition were applied to the study of actual behaviour but Cullen (1976) is surely right when he suggests that behaviour is seen as a mechanistic response to social and spatial structures. It is seen as action space or in terms of place utility. These concepts, deriving from Wolpert (1965), have, however, provided a useful starting point for the study of residential mobility.

Residential mobility, or more broadly housing demand, was identified by Robson (1973) as a potential illuminator of the structure of residential areas. Robson (1975) himself modelled the relationship between individual household characteristics and the aggregate ecology of the town, while the field has been adequately reviewed by Johnston (1971) and Short (1978a). Initially studies examined the relationships between behavioural attributes, particular stage in the life-cycle, and housing characteristics (e.g. Herbert, 1973), but the match of household to housing market has never been as clear-cut in Britain as in the US. Housing availability, access to mortgages and provision of public housing were also significant factors in shaping housing demand and influencing residential mobility (Morgan, 1976; Lewis, 1982). In other words, it was not just a matter of a household choosing or preferring a particular house-type or move, there were other forces that constrained their behaviour. Individuals do exercise choices and do have preferences which may be revealed in behaviour but the range of choice and reality of preference are constrained by positions in the social structure and in the spatial structure (Eyles, 1971). The significance of constraints on behaviour has been well noted in the study of residential mobility and housing in general, with contributions from, among others, Murie (1975) on rules of eligibility to public and private housing, Gray (1976) on selection procedures in public housing and Short (1978b) on the causes of residential mobility. In general terms, 'constraint theory' has led to considerations of class and power and, in housing, to analyses of housing supply and the role of gatekeepers. But such considerations take us well beyond the behavioural.

Relevance, Problems and a Theoretical Extravaganza

Interest in constraints co-incided with and was part of a broader based debate in the academic world. Constraints imply limits to action and to access to resources which in turn suggest the existence of social problems. These themes were highlighted by the 'relevance movement' of the early 1970s which seemed to herald the incorporation of yet more American concerns. But there is more to it than that. The context of the movement may be said to involve the liberalism of the Kennedy era and the liberalisation of social and political affairs of the Macmillan and Wilson years. The apparent universal affluence as well as the optimism and establishment complacency of the 1960s engineered a state of affairs that helped lead to discontent and highlight disadvantage. Academic disciplines sought to incorporate these concerns with studies by Harrington (1962), Galbraith (1958) and Coates and Silburn (1970) being important. Of course, interest in discontent and disadvantage never dominated the social sciences, including social geography. But these concerns, exerting an explicit influence on social geography and resulting in the dilution of the spatial dimension, led to 'new' themes (or more accurately a rediscovery of older ones, cf Reclus, Fleure).

First discussed in the US, the pages of Area were used to explore the relevance movement, which took the form of a discussion between 'liberals' and 'radicals', the former regarding social problems as being systemic malfunctions capable of solution through political and administrative action (see Berry, 1972) and the latter seeing them as being produced by contradictions inherent in the mode of social organisation and production (although elsewhere see Harvey, 1974). This schism, if it can be termed that as the significant philosophical and political differences arose from interests in different types and levels of research problem, helped to create policy (reformist) and theory (radical) dimensions in social geography. This description is, in some ways, a disservice to all: reformists used theory, especially from welfare economics and pluralist political science; radicals were interested in the nature and instigation of change (and hence policy); and the bulk of social geographers continued with the traditional academic tasks of pattern identification and analysis of process. What the relevance debate did, however, was to bring to a prominent position problem-oriented variables as inputs for such analyses (see, as an example, Herbert and Smith, 1979) and, eventually, a policy-oriented focus (see Kirby et al, 1984).

But the geographical investigation of certain social issues and problems predates the relevance debate. Race and segregation, for example, have been investigated within the spatial scientific and behavioural traditions. Jones (1979) has long been interested in examining the changing distributions of ethnic minorities in British cities. So too has Peach (1968, 1975; also Peach and Shah, 1980) although his interests also focus on the areas of origin of migrants, the migration process and migrant adaptation. Boal (1976) has examined the behavioural parameters of ethnic residential segregation as well as the effects of constraints on behaviour. Increasingly, however, there has been a focus on the broader social and economic processes within which race and segregation are located. Thus Boal (1972) has employed a conflict model of society to examine ethnic social relations in the city rather than the implied consensual model of the spatial scientific and behavioural perspectives (see Eyles, 1974). Lee (1977) has examined the relationship between race and income in London, while Doherty (1973) has explicitly analysed race and class and economic structure.

Other social problems have received geographical treatment; for example, crime (Herbert, 1982), education (Kirby, 1979) and medical care (Knox, 1978). Most such studies have located the spatial pattern in broad societal processes. But the process that has been most fully examined is that of accessibility, through which constraints on resource distributions and the problems of particular social groups can be explored. Thus, for example, accessibility has been defined in geographical (journey to a service) and in social (the impact of age, gender and class) terms (see Moseley 1979; Whitelegg, 1982). Analyses have also been undertaken of the elderly and mothers with young children and their access problems. In housing, analysis of constraints have centred on the behaviour of decision-makers, acting as gatekeepers, controlling access to the scarce resource of desirable housing (Rex & Moore, 1968; Pahl, 1975). The centrality of these gatekeepers - the managers and controllers of the housing market - led to the perspective being labelled the managerialist thesis. It was extensively criticised for ignoring the context of housing decision, namely financial and investment policy (Duncan, 1977), although its later versions do contain notions of power relations (Williams, 1978). But most 'accessibility' studies have failed to answer the telling criticism of Lee (1979) concerning the manner in which problems arise and are defined, i.e. what is a social problem? With such a view, the social problems or 'liberal' perspective itself becomes problematic. If it is analysing problems incapable of solution within existing frameworks, then it merely becomes another exercise in pattern identification and its normative discussions are, as Berry suggested, so much bleeding heart liberalism.

Such a view may be regarded as unfair given the prominence of considerations of equity and justice in the synthesising work of the social problems perspective, welfare geography. Initiated by the British economic geographer, Smith (1973), in his investigations of the US, it soon became a leading theme in British social geography. Smith (1977; 1979) extended his analyses to British regions and cities (as well as world-wide examples) and inspired other levels of living studies (e.g. Knox, 1975). It is intriguing, however, that although normative issues were made explicit in the welfare approach, it is perhaps the means of study rather than its end that seemed dominant. Welfare geography is thus regarded as part of the 'social indicators movement' in which the monitoring of social conditions occurs to effect ameliorative policy. Because of the social and conceptual distance between policy-makers and academics, the social indicators approach has become technicalist rather than normative in emphasis. The approach has assisted in the emergence of an interdisciplinary perspective and its moral stance is not in doubt in that it has identified the deprived areas of town and country (e.g. Boal et al, 1978; Coulter, 1978; Shaw, 1979). But its terms of reference remain similar to the ecological perspective in its pattern identification. The welfare approach may be largely regarded as a regional geography which emphasises social differences and inequalities.

This is not to say that the welfare approach has not addressed issues of power and contradiction. Kirby (1982), for example, sees the perspective as a starting point for analysing the politics of location. Building on the work of Smith (1977) and Taylor and Johnston (1979), he suggests that local conflicts arise from externality effects and that there exist spatial cleavages within the planning process. Further, Knox and Cottam (1981) argue that both urban and rural deprivation can be seen as outcomes of the development of late capitalism. Such comments are important but it does not always appear that welfare-oriented investigations have been fully located within such a structural context. The legacy of the welfare approach has in fact been an empirical extravaganza with theoretical and structural consideratons as secondary. It has greatly added to our knowledge of man's inhumanity to man at a variety of spatial scales (see Coates, et al, 1977) and in different locales (Knox, 1982; Phillips and Williams, 1984). The extravaganza has, therefore, a unitary rather than eclectic appearance and theoretical development has largely fossilised. Theoretical developments have in fact stemmed mainly from the radical perspective, which merged with broader considerations into a philosophical debate. The outcome of this debate has been a theoretical extravaganza and eclecticism, the ideas of which have been put forward to enhance explanatory power. Throughout the 1970s, the debate raged with the elaboration of the marxist debate in geography, parallel developments in humanistic theorising emanating mainly from the US but with British- and Irish-born geographers (e.g. Relph, Buttimer) to the fore, the questioning of the ideological bases of geographical perspectives (see Gregory, 1978) and the emergence of a holistic approach to social science often termed political economy. Such interests meant that a distinctive geographical dimension seemed increasingly difficult to justify. They also meant that there was an underemphasis of the empirical with the major theory-informed studies of the nature of society emerging from other disciplines (e.g. Townsend (1979) on poverty and deprivation and Gough (1979) and Offe (1984) on welfare systems). This period of social geography was also critical, indeed almost inimical, to the policy process.

In Britain, the most influential set of ideas was that of structuralist marxism. Drawing on the ideas of Castells (1977) and other continental marxists, there has been much discussion on the nature of housing provision and urban structure (see Harloe, 1975). The discussion has been largely theoretical, although some studies of British housing policy have been carried out which regard the policy as a reflection of working class pressures, a reconcilation of different capitalist interests and an assistance in preserving the stability of the social system (e.g. Boddy, 1976; Dickens, 1977). While far more empirical work has been carried out in the US, the marxist perspective as utilised in the UK has begun to break down the barriers between economic and social geography. Industrial restructuring and the emerging territorial divisions of labour are seen as being clearly associated with the unemployment and social problems of, say, the inner cities (see Massey and Meegan, 1978; Thrift, 1979; Massey, 1984).

The marxist perspective has addressed questions about the nature of the economic system and the role of the state in late capitalism. Its specific analyses are beginning to point to the limitations of very general explanatory theories. Thus, the underconsumptionist explanations of residential differentiation and suburbanisation, deriving from the US, are shown not to fit the British experience (see Bassett and Short, 1980). The state is seen as needing to legitimise a particular economic order and help maintain the conditions for the accumulation of capital but it must also respond to other pressures. Working class friendly societies and trade unions (pressures from 'below') helped shape national insurance and health service (Eyles and Woods, 1983). Urban social movements may develop to challenge particular geographical and economic arrangements. Mainly studied in France, Spain and the US (Piven and Cloward, 1977; Castells, 1978; Weiner, 1981), such movements demonstrate the importance of organisation and of consciousness, a collective manifestation of which is necessary to challenge the existing scheme of things. This consciousness may be a localised phenomenon which challenges national forces (see Duncan and Goodwin, 1982). But in the context of perspectives, it points to the unsatisfactory model of man deriving from the structuralist marxist approach. The idea of man as a conscious, creative force has been pointed to in both British and American social geography (Eyles, 1981; Duncan and Ley, 1982). It also allows our attention to focus specifically on humanistic perspectives.

Developing along a track almost parallel to the marxist perspective, humanistic geography has utilised the ideas of phenomenology, existentialism, symbolic interactionism, interpretative sociology and interpretative anthropology. This utilisation has been seen as necessary to address the question of how better may the researcher come to understand the world as it is seen and used by lay actors. So far most analyses have been theoretical, contemplative, introspective or literary (see Pocock, 1981; Gold and Burgess, 1982). Its eclecticism has been utilised by Jackson and Smith (1984) to explore the whole field of social geography. They adopt "an interactionist perspective which aims to uncover how social structure is defined and maintained through social interaction, and which studies how social life is constituted geographically through the spatial structure of social relations" (Jackson and Smith, 1984, vii), but in this task, the strength of eclecticism is not fully demonstrated with most examples being taken from studies of segregation: and perhaps until the purpose of the broad sweep of ideas is demonstrated, eclecticism must be considered a weakness, i.e. the relations between the ideas require specification.

The work of Jackson and Smith does, however, point to the developing convergence of different sets of ideas to enhance explanatory power. It is rather a search for philosophically acceptable common ground between what have been considered competing perspectives, i.e. between marxism and humanism and between marxist and weberian approaches to the study of social structure. These searches are in part explored by Johnston (1982) and have utilised critical theory and the theory of structuration. In other words, they have involved an examination of social theory. Thus, the critical theory of Habermas has been used to point to a critical geography of the future (Gregory, 1978) and utilised to assess the interrelations between state, economy and everyday life and the development of economic, political and legitimisng crises in the health field (Eyles et al 1982; Eyles and Woods, 1983). The theory of structuration (Giddens, 1979; 1981) has much in common with the interactionist perspective outlined above. It points to the mutual relation of human agency and structure in everyday practices. Structures are created by people but in turn constrain their activities. The process is further complicated by structures being open to modification by human actions. Structures can also enable certain activities to take place. Gregory (1981) has advocated the incorporation of 'human agency' into social geography and Giddens has argued for the inclusion of geographical ideas into social theory. Although his recent comments on both geography and history seem trite (Giddens, 1984), his work is important for the insights it provides into agency and structure in everyday practice. Similar insights have been achieved by anthropological and journalistic accounts of life at the social geographical level of analysis - the locality and neighbourhood - and which imply a contemporary significance of early notions, namely locality consciousness (see Seabrook 1982; 1983; Wallman, 1984). If consciousness is locality based, it may provide support for the view that space and location are vital for the analysis of social relations (see also Urry, 1981a; b) and challenge Saunders' (1981) view of its marginality.

The relative significance of agency (the essence of experience) and structure (rules and resources organised as social practices) will perhaps depend on the scale of analysis. Structural analysis remains attractive at the level of the city (Johnston, 1980) and the state (Johnston, 1981). Studies of everyday experience have utilised the work of the weberian-influenced Parkin (1979) with his notions of social closure, exclusion and prestige influencing studies of segregation (Philips, 1981). This experience has not as yet been the focus of many social geographic investigations in Britain, although there is an emerging interest in the perceptions and experiences of different groups to health and health care (Cornwell, 1984, Donovan, 1984). Such studies also demonstrate the structural parameters within which such experiences occur.

Empirical investigations have yet to 'catch up' with the theoretical eclecticism. Indeed, the latter continues apace with, for example, the 'rethinking of behavioural geography' (Cox & Golledge, 1981). A reason why there has been less than necessary empirical support has been the lack of necessary methodological developments. It is an irony that the theoretical and methodological eclecticisms have not only been divorced but have never been in the same church, with the former being in part instigated because of the spatial scientific monopoly of the latter. But methodological developments are now being given fuller consideration. A realist perspective on method is provided by Sayer (1984). Eyles (1985), basing his work on the three interrelated but irreducible knowledge interests identified by Habermas (1972), has utilised self-reflection, empirical scientific analyses and hermeneutics (or interpretation) to identify and discusses in a theory-informed framework senses of place and their relations to everyday life in a small English town.

If these methods are derived from philosophical and theoretical deliberations, other suggestions are more 'empirically-grounded'. The significance of consciousness and the relation of agency and the structure have led to the introduction of qualititative methods usually employed in sociology and social anthropology. Case-studies and analytic induction (see Mitchell, 1983) have been used to investigate the working lives and health experiences of East Londoners (Cornwell, 1984). Participant observation has been utilised and assessed by Jackson (1983). While being nothing new (see Le Play and the Chicago ethnographers), this methodological concern again reflects what is happening in the other social sciences (Burgess, 1982; 1984) where the limitations of empiricist enquiry and its monopoly of quantitative methods are being assessed. Indeed, it might be possible to suggest that qualitative methodologies are the 'new revolution' for human geography in general, as the effects of economic crisis, public expenditure cuts and further relegation of higher education in public and government esteem have meant the forceful resurfacing of policy-relevant, problem-oriented spatial science. In such a climate, method is likely to be more important than theory and through it different ways of looking at and gathering information about the world may be highlighted. It is through method that the issues of theory and explanation may have to be explored.

Conclusions

With anthropological method, it would appear that we have come full circle as such was a way of looking at the world in social geography in the 1930s. Perhaps no direct parallel should be drawn because it would do a disservice to the nature and context of ideas. In the intervening years, however, the subject in Britain has been dominated by eclecticism. But it is an eclecticism of different forms at different times and by different groups of scholars. While there has been an underlying theme of a geography of society, there has also occurred an eclecticism of perspectives and topics which has made synthesising the discipline or difficult task. Jones and Eyles (1977) thus fall back on scientific ordering, concepts, pattern and process, while Knox (1982) covers a vast array of topics without a strong, unifying theme. It has also made difficult policy contributions as the same theme, e.g. the inner city (Hall, 1981), has attracted different interpretations and approaches. There has also been methodological eclecticism, particularly in the spatial science framework. This has led to a refinement of technique and a broadening of input variables without confronting the abstracted natures of the methods themselves. Social geography has also been characterised by theoretical or conceptual eclecticism with the aim of enhancing explanatory power with a resultant diminution of a geographical dimension.

Eclecticism has the purposes of improving analysis and explanation but with certain costs. The major cost must be the problem of the contribution of 'geography' in social geography. While the discipline must be part of an integrated social science, it must recognise and argue for the spatial dimension as it looks at society-in-environment. Within this integrated framework, its strengths must come from its distinctive perspective, its methods and its theories. To emphasise these points, we may return to the beginning of the subject. Thus Schluter (1899) argued that anthropogeography is concerned with the form, shape and spatial arrangement or association of visible and observable geophenomena (see Berry and Kasarda, 1977). It has complete freedom as to how it explains the interactive relation between man and his environment. Hettner (1927) noted that geography's unity comes from its method with the goal of the chorological point of view being to know the character of regions and places through understanding different realms and manifestations of reality (see Hartshorne, 1959). While this is not the place to replay the debates beween chorology and spatial analysis, the idiographic and the nomothetic, or essence and appearance, the keys to modern social geography may be found in these views. The unity of social geography and of the whole discipline stem from their theory-informed methodologies as they interrogate the society-environment relation with due (but not undue) regard to space and from their theoretical coherence as they attempt to understand and explain the nature, forms and context of social relations in specific environing conditions. Methods, theories and explanatory schemes may be invented, begged, stolen or borrowed but it must be this vigorous empirical application in specific historical and contemporary contexts, by which a subject is judged. For British social geography, the jury is still out.

References

Agnew, J.A. & J.D. Duncan	1981	The transfer of ideas into Anglo-American geography, Progress in Human Geography 5, 42-57
Bassett, K. & J. Short	1980	Housing and residential structure. RKP: London
Berry, B.J.L.	1972	More on relevance and policy analysis, Area. 4, 77-80
Berry, B.J.L. & J.D. Kasarda	1977	Contemporary urban ecology. Macmillan: NY
Boal, F.W.	1969	Territoriality on the Shankill-Falls divide, Irish Geography 6, 30-50
Boal, F.W.	1972	The urban residential sub-community, Area 4, 161-8
Boal, F.W.	1976	Ethnic residential segregation, in D.T. Herbert and R.J. Johnston (eds) Social areas in cities, vol. 2. Wiley: Chichester
Boal, F.W. et al	1978	Social problems in the Belfast urban area, Queen Mary College,

		Department of Geography, Occasional Paper 12
Boddy, M.	1976	The structure of mortgage finance, Trans. Institute of British Geographers 1, 20-33
Brunhes, J.	1920	Human geography. Rand-McNally: Chicago
Burgess, R.	1982	Field research. Allen & Unwin: London
Burgess, R.	1984	In the field. Allen & Unwin: London
Castells, M.	1977	The urban question. Arnold: London
Castells, M.	1978	City, class and power. Macmillan: London
Coates, B.E. et al	1977	Geography and inequality. Oxford UP: London
Coates, K. & R. Silburn	1970	Poverty: the forgotten Englishmen. Penguin: Harmondsworth
Cohen, A. (ed)	1982	Belonging. Manchester UP: Manchester
Community Development Project (CDP)		1975 Coventry and Hillfields (2 vols). CDP: Coventry
Connell, J.	1974	The metropolitan fringe, in J.H. Johnston (ed) Suburban growth. Wiley: Chichester
Cornwell, J.	1984	Hard-earned lives. Tavistock: London
Coulter, J.	1978	Grid-square census data as a source for the study of deprivation in British conurbations, University of Durham, Department of Geography, Census Research Unit, Working Paper 13
Cox, K. & R.G. Golledge (eds)	1981	Behavioural problems in geography revisited. Methuen: London
Cullen, I.G.	1976	Human Geography, regional science and the study of individual behaviour, Environment and

		Planning A 8, 397-409
Dickens, P.	1977	Social change, housing and the state, in M. Harloe (ed) Second conference on urban change and conflict. CES: London
Dickinson, E.	1957	The west European city. RKP: London
Doherty, J.	1973	Race, class and residential segregation in Britain, Antipode 3, 45-51
Donovan, J.	1984	Black people's health: a different approach? Paper presented to ethnicity and health conference, Leeds
Downs, R.M.	1970	The cognitive structure of an urban shopping centre, Environment and Behaviour 2, 13-39
Dunbar, G.S.	1977	Some early occurrences of the term 'social geography', Scott, Geog. Mag. 93, 15-20
Dunbar, G.S.	1981	Elisee Reclus, an anarchist geographer, in Stoddart op. cit.
Duncan, J. & D. Ley	1982	Structural marxism and human geography, Annals Assoc. Am. Geog. 72, 30-59
Duncan, S.S.	1977	The housing question and the structure of the housing market, Journal of Social Policy 6, 385-412
Duncan, S.S. & M. Goodwin	1982	The local state and restructuring social relations, Int. J. Urban Reg. Res. 6, 157-86
Evans, E.	1942	Irish heritage. Dundalgon: Dundalk
Evans, E.	1951	Mourne country. Dundalgon: Dundalk
Eyles, J.	1971	Pouring old sentiments into new theories, Area 3, 242-50
Eyles, J.	1974	Social theory and

		social geography, *Progress in Geography* 6, 22-87
Eyles, J.	1981	Why geography cannot be marxist, *Environment & Planning A* 13, 1371-88
Eyles, J.	1984	The examination of social phenomena in space, *Geojournal* 9, 247-56
Eyles, J.	1985	*Senses of place*. Silverbrook Press: Warrington
Eyles, J. & D.M. Smith	1978	Social geography, *American Behavioural Scientist* 22, 41-58
Eyles, J. & K.J. Woods	1983	*The social geography of medicine and health*. Croom Helm: London
Eyles, J. et al	1982	Spatial resource allocation and state practice, *Regional Studies* 16, 239-53
Febvre, L.	1925	*A geographical introduction to history*. Knox: NY
Firey, W.	1947	*Land use in central Boston*. MIT Press: Cambridge, Mass
Fitzgerald, W.	1946	Geography and its components *Geographical Journal* 197, 272-3
Fleure, H.J.	1919	Human regions, *Scott. Geogr. Mag*. 35, 94-105
Fleure, H.J.	1929	*An introduction to geography*. Benn: London
Fleure, H.J. & T.C. James	1916	Geographical distribution of anthropometric types in Wales, *J. Royal Anthrop. Inst. GB Ireland* 46, 35-153
Forde, C.D.	1934	*Man, habitat and society*. Methuen: London
Freeman, T.W.	1965	*One hundred years of geography*. Duckworth: London
Galbraith, J.K.	1958	*Affluent society*. Hamish Hamilton: London

Gans, H.J.	1962	Urbanism and suburbanismm as ways of life, in A. Rose (ed) Human behaviour and social processes RKP: London
Giddens, A.	1979	Central problems in social theory. Macmillan: London
Giddens, A.	1981	Contemporary critique of historical materialism. Macmillan: London
Giddens, A.	1984	The constitution of society. Polity Press: Oxford
Glass, R.	1948	The social background of a plan. RKP: London
Glass, R.	1955	Urban sociology in Great Britain, Current Sociology 4, 5-19
Goddard, J.B.	1978	The location of non-manufacturing activities within manufacturing industries, in F.E.I. Hamilton (ed) Contemporary industrialisation. Longman: London
Gough, I.	1979	The political economy of the welfare state. Macmillan: London
Gold, J.	1980	An introduction to behavioural geography. Oxford UP: London
Gold, J. & J. Burgess (eds)	1982	Valued environments. Allen and Unwin: London
Gould, P. & R. White	1968	The mental maps of the British school leavers, Regional Studies 2, 161-82
Gray, F.	1976	Selection and allocation in council housing. Trans. Inst. Brit. Geogr. 1, 34-46
Gregory, D.	1978	Ideology, science and human geography. Hutchinson: London
Gregory, D.	1981	Human agency and human geography, Trans. Inst.

Brit. Geogr. 6, 1-18

Habermas, J. 1972 Knowledge and human interests. Hutchinson: London

Hall, P. 1981 Inner city in context. Heinemann: London

Hannerz, U. 1980 Exploring the city. Columbia UP: NY

Harloe, M. (ed) 1975 Captive cities. Arnold: London

Harrington, M. 1962 The Other America. Penguin: NY

Harrison, J. & P. Sarre 1971 Personal construct theory in the measurement of environmental images, Environment and Behaviour 3, 31-74

Harrison Church, R.J. 1957 The French school of geography, in Taylor op. cit.

Hartshorne, R. 1959 Perspective on the nature of geography. Rand-McNally: NY

Harvey, D.W. 1973 Social justice and the city. Arnold: London

Harvey, D.W. 1974 What kind of geography for what kind of policy? Trans. Inst. Brit. Geogr. 63, 18-24

Harvey, D.W. 1982 Limits to capital. Blackwell: Oxford

Herbert, D.T. 1967 Social area analysis: A British study, Urban Studies 4, 41-60

Herbert, D.T. 1973 The residential mobility process, Area 5, 44-8

Herbert, D.T. 1975 Urban neighbourhoods and social geographical research, in A.D.M. Phillips & B.J. Turton (eds) Environment, man and economic change. Longman: London

Herbert, D.T. 1982 The geography of urban crime. Longman: London

Herbert, D.T. & D.M. Smith (eds) 1979 Social problems and the city. Oxford UP: London

Herbert, D.T. & 1982 Urban geography.

C.J. Thomas		Wiley: Chichester
Hettner, A.	1927	*Die Geographie*. Hirt: Breslau
Hoke, G.W.	1907	The study of social geography, *Geographical Journal* 29, 64-7
Hudson, R.	1974	Images of the retailing environment, *Environment and Behaviour* 6, 470-94
Hudson, R.	1976	Linkage studies of the individual with models of aggregate behaviour, *Trans. Inst. Brit. Geogr*. 1, 159-74
Jackson, P.	1983	Principles and problems of participant observation, *Geografiska Annaler* 65B, 39-46
Jackson, P. & S. Smith (eds)	1981	*Social interaction and ethnic segregation*. Academic Press: London
Jackson, P. & S. Smith	1984	*Exploring social geography*. Allen & Unwin: London
Jenkins, D. et al	1960	*Welsh rural communities*. University of Wales Press: Cardiff
Johnston, R.J.	1971	*Urban residential patterns*. Bell: London
Johnston, R.J.	1976	Residential area characteristics, in D.T. Herbert and R.J. Johnston (eds) *Social areas in cities*, vol. 1. Wiley: Chichester
Johnston, R.J.	1980	*City and society*. Penguin: Harmondsworth
Johnston, R.J.	1981	The state and the study of social geography, in Jackson and Smith *op. cit*.
Johnston, R.J.	1982	*Philosophy and human geography*. Arnold: London
Jones, E.	1960	*Social geography of Belfast*. Oxford UP: London
Jones, E.	1980	Social geography, in E. Brown (ed) *Geography:*

		yesterday and tomorrow. Oxford UP: London
Jones, E. & J. Eyles	1977	An introduction to social geography Oxford UP: Oxford
Jones, P.N.	1979	Ethnic areas in cities, in Herbert and Smith op. cit.
Kinvig, R.H.	1953	The geographer as humanist, Advancement of Science 10, 157-68
Kirby, A.M.	1979	Education, health and housing. Saxon House: Farnborough
Kirby, A.M.	1982	The politics of location. Methuen: London
Kirby, A.M. et al (eds)	1984	Public service provision and urban development. Croom Helm: London
Kirk, W.	1951	Historical geography and the concept of the behavioural environment, Indian Geog. Journal 25, 152-60
Kirk, W.	1963	Problems of geography, Geography 48, 357-71
Knox, P.L.	1975	Social well-being. Oxford UP: London
Knox, P.L.	1978	The intraurban ecology of primary medical care, Environment and Planning A 10, 415-34
Knox, P.L.	1982	Urban social geography. Longman: London
Knox, P.L. & M.B. Cottam	1981	Rural deprivation in Scotland, TESG 72, 162-75
Lee, R.	1979	The economic basis of social problems in the city, in Herbert and Smith op. cit.
Lee, T.	1977	Race and residence. Oxford UP: Oxford
Lewis, G.J.	1982	Human migration. Croom Helm: London
Lewis, I.M.	1976	Social anthropology in perspective. Penguin: Harmondsworth
Ley, D.	1983	A social geography of the city. Harper and

Britain

Row: NY

Lloyd, P. & P. Dicken 1972 Location in space. Harper and Row: London

McKenzie, R.D. 1924 The ecological approach to the study of the human community, American J. Sociol. 30, 287-301

Massey, D.B. 1984 Spatial divisions of labour. Macmillan: London

Massey, D.B. & R.A. Meegan 1978 Industrial restructuring versus the cities, Urban Studies 15, 273-88

Mitchell, S.C. 1983 Case and situational analysis. Sociol. Review 31, 187-211

Mogey, S.M. 1956 Family and neighbourhood. Oxford UP: Oxford

Morgan, B.S. 1976 The basis of family status segregation, Trans. Inst. Brit. Geogr. 1, 83-107

Moseley, M. 1979 Accessibility. Hutchinson: London

Moser, C.A. & W. Scott 1961 British towns. CUS: London

Murie, A. 1975 Household movement and housing, University of Birmingham, Centre for Urban & Regional Studies, Occasional Paper 28

Newby, H. 1977 The deferential worker. Allen Lane: London

Offe, C. 1984 Contradictions of the welfare state. Hutchinson: London

Pahl, R.E. 1965a Trends in social geography, in P.J. Chorley and P. Haggett (eds) Frontiers in geographical teaching. Methuen: London

Pahl, R.E. 1965b Urbs in rure, London School of Economics, Monographs in Geography 2

Pahl, R.E. 1967 Sociological models in

		geography, in R.J. Chorley and P. Haggett (eds) *Models in geography*. Methuen: London
Pahl, R.E.	1968	A perspective in urban sociology, in R.E. Pahl (ed) *Reading in urban sociology*. Pergamon: Oxford
Pahl, R.E.	1975	*Whose city?* Penguin: Harmondsworth
Park, R.E.	1936	Human ecology, *American J. Sociol*. 42, 1-15
Parkin, F.	1979	*Marxism and class theory*. Tavistock: London
Peach, G.C.K. (ed)	1968	*West Indian migration in Britain*. Oxford UP: London
Peach, G.C.K. (ed)	1975	*Urban social segregation*. Longman: London
Peach. G.C.K.	1981	Conflicting interpretations of segregation, in Jackson and Smith *op. cit*.
Peach, G.C.K. & S. Shah	1980	The contribution of council house allocation to West Indian desegregation in London, 1961-71, *Urban Studies* 17, 333-41
Peate, I.C. (ed)	1930	*Studies in regional consciousness and environment*. Oxford UP: London
Peet, R. (ed)	1977	*Radical geography*. Maaroufa Press: Chicago
Phillips, D.	1981	The social and spatial segregation of Asians in Leicester, in Jackson and Smith *op. cit*.
Phillips, D.R. & A. Williams	1984	*Rural Britain*. Blackwell: Oxford
Piven, F. & R.A. Cloward	1977	*Poor people's movements*. Pantheon: NY
Pocock, D. (ed)	1981	*Geography and literature*. Croom Helm: London

Britain

Pocock, D. & R. Hudson	1978	Images of the urban environment. Macmillan: London
Raban, J.	1975	Soft city. Fontana: London
Rees, A.	1950	Life in a Welsh Countryside. University of Wales Press: Cardiff
Rees, P.H.	1970	Concepts of social space, in B.J.L. Berry and F.E. Horton (eds) Geographic perspectives on urban systems. Prentice-Hall: Englewood Cliffs, NJ
Reissman, L.	1964	The urban process. Free Press: Glencoe
Rex, J. & R. Moore	1968	Race, community and conflict. Oxford UP: London
Robson, B.T.	1969	Urban analysis. Cambridge UP: Cambridge
Robson, B.T.	1973	Introduction, in B.D. Clark and M.B. Gleave (eds) Social patterns in cities. IBG: London
Robson, B.T.	1975	Urban social areas. Oxford UP: London
Robson, B.T.	1981	Geography and social science: the role of Patrick Geddes, in Stoddart op. cit.
Roxby, P.M.	1925	The distribution of population in China, Geogr. Review 15, 1-24
Roxby, P.M.	1930	The scope and aims of human geography, Scott. Geogr. Mag. 46, 276-90
Saunders, P.	1981	Social theory and the urban question. Hutchinson: London
Sayer, A.	1984	Method in social science. Hutchinson: London
Schluter, O.	1899	Bemerkungen zur Siedungsgeographie, Geographische Zeitschrift 5, 65-84
Seabrook, J.	1982	Unemployment. Paladin: St. Albans

Seabrook, J.	1983	The idea of neighbourhood. Pluto Press: London
Shaw, J.M. (ed)	1979	Rural deprivation and planning. GeoBooks: Norwich
Short, J.	1978a	Residential mobility, Progress in Human Geography 2, 419-47
Short, J.	1978b	Residential mobility in the private housing market in Bristol, Trans. Inst. Brit. Geogr. 3, 533-47
Smith, D.M.	1973	Geography of social well-being in the US. McGraw-Hill: NY
Smith, D.M.	1977	Human geography: a welfare approach. Arnold: London
Smith, D.M.	1979	Where the grass is greener. Penguin: Harmondsworth
Sorokin, P.A.	1964	Frederic Le Play's school, in F.S. Chapin (ed) Contemporary sociological theories. Harper and Row: NY
Spencer, C. & M. Weelman	1981	The microgenesis of cognitive maps, Trans. Inst. Brit. Geogr. 6, 375-84
Spencer, D.	1973	An evaluation of cognitive mapping in neighbourhood perception, University of Birmingham, Centre for Urban and Regional Studies, Memorandum 23
Stacey, M.	1969	The myth of community studies, Brit. J. Sociol. 20, 134-47
Strathern, M.	1982	The village as an idea, in Cohen op. cit.
Stoddart, D.R. (ed)	1981	Geography, ideology and social concern. Blackwell: Oxford
Tatham, G.	1957	Environmentalism and possibilism, in Taylor op. cit.
Taylor, G.	1946	Urban geography. Methuen: London

Britain

Taylor, G. (ed)	1957	Geography in the twentieth century. Methuen: London
Thrift, N.	1979	Unemployment in the inner city, in D.T. Herbert and R.J. Johnston (ed) Geography and the urban environment, vol. 2. Wiley: Chichester
Timms, D.W.G.	1971	The urban mosaic. Cambridge UP: Cambridge
Townsend, P.	1979	Poverty in the United Kingdom. Penguin: Harmondsworth
Urry, J.	1981a	The anatomy of capitalist societies. Macmillan: London
Urry, J.	1981b	Localities, regions and social class, Int. J. Urban Reg. Res. 5, 455-74
Vidal de la Blache, P-H.	1913	Des caractères distinctifs de la geographie, Annales de Géographie 22, 289-99
Wallman, S.	1984	Eight London households. Tavistock: London
Walmsley, D.J. & G.J. Lewis	1984	Human geography: behavioural approaches. Longman: London
Watson, J.W.	1957	The sociological aspects of geography, in Taylor op. cit.
Weiner, R.	1981	Cultural marxism and political sociology. Sage: Beverly Hills
Whitelegg, J.	1982	Inequalities in health care. Straw Barnes: Retford
Williams, P.	1978	Urban managerialism, Area 10, 236-40
Williams, W.M.	1960	The sociology of an English village. RKP: London
Williams, W.M.	1963	A west country village. RKP: London
Wirth, L.	1938	Urbanism as a way of life, American J. Sociol. 44, 1-24

Chapter Four

SWEDISH SOCIAL GEOGRAPHY

Sven Godlund

"Though this be madness, yet there is method in it" (Shakespeare)

INTRODUCTION AND FUNDAMENTAL ASPECTS

In this contribution I have used the same intellectual approach as that adopted by the American philosopher Nelson Goodman, who, in his book "Ways of Wordmaking" (1978), declares that it is impossible to state "what is art?"; what you can say is "when is art?". This statement has led me to an unbounded search for the changing contexts within the broad sector of social geography and to express these findings as examples, chosen and analysed on the basis of my paradigm; the meaning of the word paradigm corresponds here to that of Kuhn and the development of this approach by Tornebohm.

Inasmuch as this essay is based on extracts from a larger work it has been possible to follow not only the long-range trends within what can be termed "social geography" but also to give some insight as regards general changes in geography as a totality, obviously with special emphasis on changes in Sweden, which, incidentally, are mainly the same as in the other Nordic countries.

In this abbreviated version I have used what we can call a continual process model, where the changing ideas, aspects etc. of the subjects studied take the shape of a relay race. I have derived the idea of this analogy between research process and relay race from the Finno-Swedish philosopher Hans Ruin's book "The Mysticism of Poetry" (1935). By both accumulating and detaching ideas, aspects, etc, I have obtained a long-range survey of Swedish social geography. The road I have travelled has been divided into certain phases, each dominated by a characteristic view or methodology. However, this being a process study, the phases are linked together.

The relay itself is called "territory" - it could also be named space or area. This is the summarising expression indicating the common geographical idea and arsenal of aspects and theories, methodologies and methods, applicable to or derived from our knowledge and studies of society and its different systems and elements. "Territory" is here the explicit term for the research domain of geography: a common bank of knowledge at the "meta level" but at the same time changing in stages through the influence of new thoughts, methods, etc. within or outside the geographer's world of thought and activity. Here I agree with the German philosopher Martin Heidegger's ironic attitude towards those who believe that science has a soul: an academic discipline is what we can make it!

From the above it is apparent that I have not attempted to make any definition of "social geography". Every reader must, in accordance with his/her own paradigm, accept or reject the presented examples of Swedish geographical research achievements within the theme of the study. Reality is, perhaps paradoxically, relative and man irrational, which is certainly not a paradox.

I shall here only give an "adaptable field of information", in which I have identified the examples of social geography and hope that my readers will also do so: "Man's territorial relations with his fellow-creatures and with the physical and social environment both on the local and global arenas, viewed from the horizon, conditions and possibilities of ordinary people". I believe this is the same principle as that underlying Emrys Jones' and John Eyles' book "An Introduction to Social Geography" (1977).

As the examples are just that and no more, I am wary of making comments. I have presented the authors and their works as directly as I can without any particular designation or ranking. This procedure is similar to that recommended by Peter Gould in his paper "Letting the data speak for themselves" in Annals of the Association of American Geographers in 1981. It is also used by the English historian Robert Skidelsky in his book "John Maynard Keynes - Hopes Betrayed" (1983), where he lets Keynes appear through his own ideas and publications, speeches and letters, etc. I have consequently tried to avoid imposing myself between the respective author and the readers by making more or less leading comments and allusions.

Seeing that an international presentation of this nature should not be the work of one man alone, I have asked eight of my active university colleagues in Sweden and to state which five to ten scientific geographical works have meant most to them. In all, these eight friends have separately mentioned 67 titles of books and papers. However, as many titles have been mentioned twice or more the sum total amounts to 45. About two thirds of these works have been written during the past two decades. The total number of authors is 31. No study mentioned is older than from the late 1930's, which, in my opinion, signals the need of a long-range review. Of course, this enquiry has widened my own paradigm.

Terrritory - Form - Size To the Turn of the Century 1800/1900

During the times lasting approximately until the turn of the century 1800/1900 the scene was dominated by accounts concerning different continents, countries, regions and places with descriptions of their location and configuration, extent and size, for instance as regards area and population. This positivistic approach was strongly influenced by German models introduced in Sweden, partly via Denmark, and applied not only to what one may term academically orientated presentations - very often, for organisational reasons, included in the subject history - but also to the emerging great variety of school-books during the 19th century.

This phase, which quite naturally, in view of my procedural approach, is "open" backwards in time has its centre of gravity during the two or three last decades of the 19th century, when the subject geography began to attain a profile of its own in Sweden.

Within what we can call general geography the above phase is characterised above all by voyages of discovery and the mapping of the remaining unexplored regions of the planet. In view of the fact that biologists and members of the medical profession often accompanied these voyages of discovery, the accounts of the journeys may very often include a wealth of information concerning for instance plant and animal life as well as many observations regarding the people encountered in the newly discovered areas. These descriptions could refer to appearance and bodily constitution, housing conditions, food habits, clothing and tools, means of livelihood and communication. They also mapped i.a. languages and religious ideas, rites and mental distance. Interdisciplinary or multidisciplinary mapping is by no means something belonging solely to our own day and age.

In this connection it should be noted that geographers of the generation in question, who were not interested in physical geography or voyages of discovery, set out on intellectual journeys backwards in time by studying information concerning the form and peoples of the past. Together with the then still existent historical-classic education, this explains the emergence during the later decades of the 19th century of the historical-geographic theses by Hans Hugold von Schwerin, Karl Ahlenius and Sven Lonborg.

To borrow a term from a far later study by Torsten Hagerstrand in 1961 on individual achievements and technical economic innovations, these three geographers became "gate-openers", i.e. by Schwerin obtaining the first professor ship in geography at Lund University (1897-1912), with Ahlenius being awarded the equivalent chair at Uppsala University (1902-1906).

It may be added that the last decades of the 19th century also witnessed the birth of a number of organisations with the aim of promoting Swedish geography and having their own publications. The first was the Swedish Society of Anthropology and Geography, SSAG, (1877) which introduced a publication entitled "Ymer" (1880). This publication has gradually come to contain more and more purely geographical themes with both social, economic resource and, of course, physical trends. As from 1966, "Ymer" is published as a year-book, each year being devoted to a particular theme, e.g. "Environment control and landscape"(1970), "Land use and people" (1972), "Review of the geographical research in Sweden" (1979), and "Map and society" (1984).

In 1885, the Swedish Tourist Association, STF, was established for the purpose of developing and facilitating tourism and spreading knowledge of the country and its people. Since 1886 it has published a year book, successively presenting either a number of geographically, socially and historically orientated themes on the various provinces in Sweden.

As an exponent of Swedish social-geographic research during this phase I have chosen first and foremost Adolf Erik Nordenskiold's accounts of the voyage he made (1879-1880) in the North-East Passage. They contain a rich variety of information of the kind mentioned above concerning the people along the coast of the Arctic Ocean and their living conditions.

Another example is the work of Karl Ahlenius (1897) dealing with the classification of human "races". This classification is based on skin colour and other "racial" characteristics. Karl Ahlenius also gave the first exhaustive introduction to the Swedish public of the German geographer Friedrich Ratzel's intentions and thoughts, methodology and technique in his fundamental work "Antropogeographie" (1882).

The global interests of Swedish social geographers are also found in von Schwerin's paper (1892) concerning "Mohammedanism in Africa", where he suggests that the successes of Islam in relation to Christianity depend on Islam representing a comprehensive totality, religiously and socially, economically and politically.

Territory - Size - Structure. The Phase Covering the First Three Decades of the 20th Century

The period in question can be called a phase of organisational stabilisation. Geography had expanded as a special subject at all levels with a resultant great demand for teachers, particularly those with the combinations of geography with biology, history, mathematics, or physics. At that time all four Swedish universities had full professorships and departments of geography: Lund (1897), Uppsala (1902), Gothenburg (1905) and Stockholm (1929).

In the field of geographical publications, two new series appeared during this phase. The SSAG commenced with "Geografiska Annaler" (1919). From 1922 to 1978, another periodical, "Globen", was published. It contained i.e. papers about urban and social geography. The first paper in the first volume deals, significantly enough with the questions of how to define the concept of urbanism from the geographer's point of view. In Lund, the South Swedish Geographical Society, founded in 1925 began to publish "The Swedish Geographical Yearbook".

These expansions at the organisational level were, however, not followed by an equally strong geographical-scientific development. I believe this was chiefly due to the fact that, during this period, all the professors had a natural science background. This gave their work a physical rather than a social orientation, although interest was also shown in place-name geography, a later somewhat forgotten subject in the field of securing human identity, and in geopolitics.

There was, however, some social geography in this phase, e.g. an investigation about the Finnish population group in central Sweden and in adjacent parts of Norway, resulting from immigration from Finland mainly during the 17th and 18th centuries (Lonborg, 1902). This study examined place-name and ethnographic evidence, as well as religious practices, school education, the remaining language problems and the physiognomy of the people. Helger Nelson (1913) examined a mining district in central Sweden by means of a historical-geographical survey, executed in the spirit of the dominating German regional studies, "Die Landschaftskunde".

Technical advances were made, particularly in population mapping, by Sten De Geer (1919). He also determined a number of regional divisions of the country based i.a. on the different structure in each district and the dominating social and economic contacts between larger and smaller areas.

De Geer's methodology of population mapping has been continuously developed, both in Sweden and at the international level. A similar distribution map of the population in Norway was published in 1923 by the Swedish geographer Alfred Soderlund. In many countries - as in Sweden - this type of work is to-day performed by computers with digitalisation equipment.

In 1911 the Swedish geographer and labour politician Rickard Sandler published a book as a contribution to the social debate entitled "Samhallet sadant det ar", "Society in reality". This book was printed in a revised and enlarged version in 1924. The first version, from 1911, is strongly influenced by the methodology of social economy, a discipline older than geography. Consequently, this edition contains data of a mainly macro-oriented nature, considering capital owning groups. In the revised 1924 version we find much more detailed socio-geographical aspects on man's day-do-day reality, e.g. the distribution of shopping facilities in urban districts. There is thus a clear tendency from social economy about the people towards social geography with the people.

Other examples of Swedish studies of man's

dependence on his local environment are studies of John Frodin. His work on the saeter (Sw. "fabod") system in the Siljan district of Dalarna in central Sweden (1925) is especially well-known. Frodin anticipated the change of this very old system from being a necessary integrated link in the local agrarian rhythm of activity to becoming a factor of minor importance. Many saeters were to be abandoned, with either the forest taking over or urban dwellers using them as summer residences.

Swedish social geography also looked abroad during this phase. Nelson commenced with his partly extensive, partly detailed social geographic studies about the Swedish emigrants to North America and their settlements there. However, the international summary of his work was not published until 1943.

Territory - Structure - Function. The Phase of the 1930's and 1940's

An example from the early 1930's is that of De Geer, who was interested in a combination of structural and functional geography as evidenced in built-up areas, from urban morphology and local city problems to the distribution of trade, industry and population. Another main theme was the mapping of umlands by means of commercial contacts, migration, commuting and similar indicators, exemplified by studies of Stockholm, Gothenburg and Jonkoping and some smaller towns in the suburban area of Gothenburg. In Stockholm, geographers commenced a large research programme concerning inner differentiation in Stockholm, i.a. resulting in the doctoral dissertation of William-Olsson (1937).

It may be added that several investigations within the theme town-umland were made, not only in Sweden but also in the other Nordic countries. In Norway, Tore Sund presented such a dissertation about Bergen (1947), while in Finland Olavi Tuominen published a study concerning Turku (1949). In Denmark, Aage Aagesen (1949) wrote a dissertation about the development of the railways in the country, including many detailed studies, for instance concerning Esbjerg, the large port with links to i.a. the UK, built in the 1860's.

At the beginning of the 1930's, ideas of making population projections and similar forecasts appeared from the USA and the UK. De Geer was a gate-opener in Sweden within this sphere and completed a population project for the community of

Gothenburg. In 1941, this was followed by a projection into the future regarding the population of Stockholm by William-Olsson (1941). His absolute figures proved to be too low, but his basic assumption of there being a connection between the growth of employment within the public sector of the economy and the growth of the population in the capital city constitutes a good example of Swedish social geography during this period, showing how it changed from describing structures to the implementing of functions.

During this phase, the accent was initially on physical environment determinism, its aspects and methods. However, there was gradually a reaction against this attitude, finally leading to a splitting up of the institutes of geography at the Swedish universities into one professorship of physical geography and one professorship in human geography, commencing in Uppsala in 1948 and followed by Lund in 1949. The year 1948 saw the birth of the state "Swedish Council for Social Science Research", which provided economic possibilities for geography with the emphasis on human-social themes and methodological development.

The preceding phase had been a golden age for regional geographic studies. However, during the 1930's and especially during the 1940's this type of work proved to be more and more antiquated. Earlier, such studies had been applicable as models for newly discovered countries with a primitive pattern of work and settlement, where the link between man and nature was direct and vertical, but they did not suit successively industrialised and urbanised countries, such as Sweden, where increasing communications established a more complex and horizontal contact.

External pressure for a changed attitude within Swedish geography emanated partly from other subjects such as quaternary geology, climatology and limnology, partly,and to an even greater extent, from economic and social science subjects, such as economy, sociology and statistics. However, also within the field of geography there was an inclination among the young researchers, to whom I then belonged, to react against determinism, both in the physical and historical sense. In Lund, where I studied, our "bible" was the well-known book "The Nature of Geography" by Richard Hartshorne (1939).

In two large works Gerd Enequist (1943, 1946) presents a cartographic and verbal presentation in Uppsala of the economically active population, by industry, in all communities in Sweden in 1930 and 1940. These two studies made Gerd Enequist an innovator in science. They have been of great value for regional planning, for instance.

In Lund, the arrival of Edgar Kant from Estonia in 1945 was, without doubt, a source of stimulation and innovation. Thanks to Kant, Swedish geography came into serious contact with for example Walter Christaller's central place theory (1933), with the theories of the agrarian economist J.H. von Thunen who preceded him (1826) and of course with Kant's own work about the environment and organism of the Estonian town of Tartu (1926). Kant also formed the connecting link to French geographers, such as George Chabot. In 1948, Kant interestingly wrote: "Ecological approach to social problems presupposes the knowledge of the latter. Environment cannot be studied apart from the beings for whom it is the environment .... The fact that many research-workers have lacked a foothold in sociology, ethnology, economy or history has formed the Achilles-heel of the present-day social geography."

With the first signs appearing in the 1930's, Swedish geography was soon characterised by the "quantitative revolution". Geographers were then in a position to perform a number of complicated mathematical operations based on large volumes of data. It was possible to discuss not only the classical questions in geography, viz. Where? and Why?, but also more and more How? - to understand the mechanisms of society and view its various components as a more or less integrated system.

Territory - Function - System. The Phase of the 1950's and 1960's

During the 30's, as outlined above, social geography began its liberation from nature determinism with its purely historical view. Instead, it engaged social and economic sciences of all kinds. Such was the intellectual climate when geographers seriously started to study questions of migration. They occurred mostly as parts of large, general umland studies, such as those mentioned above, for instance the study concerning the town of Jonkoping. Within this more and more socially emphasised human geographical research some main currents gradually emerged, viz. innovations and diffusions, contacts and settlement structures.

Almost always the studies in question were

applications of a positivistic and quantitative point of view, this being the general trend within all research at that time. Social geography was influenced by what I would call "the Spirit of Regional Science". Methodological references to literature from the UK and the USA became frequent, e.g. Chorley, Garrison, Gould, Haggett, Harvey, Isard, Morrill, Pred and Wolpert can be mentioned.

Towards the end of the 1940's separate geographical migration studies were produced. A very early work of outstanding methodological merit was that of Torsten Hagerstrand (1947), in which he regards a rural parish in the province of Ostergotland and the individual contacts of the inhabitants with the surrounding world. It was followed by a similar study concerning the town of Simrishamn, in Scania, also by Hagerstrand (1949).

At a symposium in Lund in 1954 Hagerstrand, together with other scientists from both Lund and elsewhere, discussed their studies. The contributions to this symposium were published as a report in the series "Lund Studies in Geography" (1957). It contained Hagerstrand's paper "Migration and Area. Survey of a Sample of Swedish Migration Fields and Hypothetical Considerations on Their Genesis", a title which clearly indicates the substance and methods of the contribution. In the same report there is also an essay entitled "Mutual Relations between Migration Fields: A Circulation Analysis" by Esse Lovgren, from Uppsala, and another paper about new aspects on geographic and social patterns of migration in Finland by the Finnish geographer Reino Ajo.

Another work of a methodological nature is that of Gunnar Olsson (1965). He tests two migration models: the stochastic migration model and the deterministic migration model, and also makes analogies between central place theory and migration.

In connection with his migration studies Hagerstrand (1953) examined questons relating to contacts in general and contacts concerning innovations and diffusions in particular. He compared figures about social geographic contacts, based on facts, with theoretically composed events, using simulating models of a Monte Carlo type. As quantitative measures of these contacts, Hagerstrand utilised the number of telephone calls in a rural area with different categories by distance. He found that, in some cases, theory coincided with reality whereas in other cases there were obvious divergences, which could be explained by mental as well as by economic and physical obstacles.

Using modern quantitative and theorically based methodology, Torsten Hagerstrand was the real forerunner or the Swedish founder of the new human geography with a social foundation based on people's intentions and possibilities, places and contacts.

New themes dealing with, for instance, investigations about telephone contacts and about the spread of the television system were attached to this line of thought, e.g. by Arne Lysen (1955). He reports on the decreasing contact field from one town (Alingsas in western Sweden) but also discusses contact defects and barriers in the area studied. The geography of the television system was studied by Gunnar Tornquist (1967).

During this phase studies on commuting and mobility also became important in understanding structural changes and as indicators of localisation and administrative divisions. Olof Nordstrom (1957) examined population changes and employment problems in the so-called glass manufacturing district in the province of Smaland, and David Jarhult (1958) studied migration, commuting and labour recruiting in connection with a new ironworks in Lulea. A study for practical purposes was that of Palsson (1958), in which he dealt with school localisation and the planning of areas of accessibility. Rural socio-spatial problems relating to employment were examined by Lewan (1967).

Colonisation expansion and contraction were further geographical research themes with a migration bias. Such studies examined the problems of populating and servicing sparsely peopled regions from both theoretical and empirical viewpoints (see Byland, 1956 and 1966; Rudberg, 1957).

During the 1950's and especially the 1960's studies about urbanisation at the regional level became starting-points for theoretical geography with the aid of simulating techniques and perceptions. The achievements of the Swedish urban researcher during this phase can be roughly divided into two main groups. The first was based on outer contacts and roles in the total pattern of settlement, e.g. Claeson (1964 and 1969) and Warneryd (1968).

Already in the 1930's questions resulting to summer vacations were raised by Swedish geographers, for instance Ljungdahl (1938). Studies of tourism have also been undertaken, e.g. Frimodig (1959) about tourism in Bohuslan, a province just north of the city of Gothenburg. Nordstrom and Solveig

Mårtensson (1966) have dealt with tourism on the Swedish island of Öland in the Baltic and Aldskogius (1967, 1969) with different Swedish regions.

The second group obtained its inspiration from for instance Lewis Mumford, Kevin Lynch and Gregor Paulsson. The latter was a Swedish art historian who supervised a large work entitled "Swedish Town" (1950 and 1953) containing not only architectural and ethnological elements but also sociological and socio-geographic facts. (A retrospective study was made by i.a. Goldfield 1981). The ideas and realisation of New Towns in the UK also influenced Swedish social geography, i.a. Sven Dahl (1955).

Another exponent among Swedish geographers of this second group is Gudrun Jarbur (1964). She studie structural changes of the population in a suburban area near Gothenburg, built in the 1950's when urbanisation flourished. For her study she divided the households in the dwelling area - comprising around 200 flats - into three social groups: I Highest level of income and social status, II Middle class, III Low income and inferior education category. She studied the situation for these three groups on three different occasions, namely in 1958, 1960 and 1963, concluding that there has been a shifting of those who have remained within the selected area from a high social status towards a lower one. This is due to the fact that the tendency to move has been largest firstly among those belonging to social group I and secondly those occupying certain types of apartments.

While these studies were primarily empirical and applied contributions, theoretical thoughts in modern social geography have not been forgotten. One can mention studies of accessibility by Claeson (1964 and 1969) and economic decline (Dahlbert (1969). With the latter work on bankruptcies pointing forward to social issues, such research was based on "soft" material and individual observations. The earlier, often aggregated "hard" data and their applications were static descriptions and pure dataism, as the Canadian philosopher Mario Bunge (1967) ironically and expressively put it.

A forerunner in Swedish geography to this new or rediscovered way of looking at society and its changes was Lennart Ameen (1964) and his analyses of social landscape in terms of privately and publicly owned space in a number of Swedish towns of varying age and having different city plans and physical foundations.

Territory - System - Action. The Phase of the 1970's and up to the Present Day

This phase commenced with what one might call a continuing outer consolidation of human geography as a scientific discipline, equal to i.a. economic and social science both from a theoretical point of view and in an applied sense. In connection with the 300th anniversary of the Central Bank of Sweden, the bank established a foundation with the aim of sponsoring scientific investigations dealing with the conditions and future of man. The departments of geography in Lund and Gothenburg received a joint grant for a project entitled "The Process of Urbanisation". The first results of this initative began to be published during the initial years of the 1970's.

The so-called "students' revolt in 1968 had only a slight effect on geographical research and education in Sweden, but the movement or rather commotion did give rise to a number of reflections and considerations within the discipline as to the role of the geographer in society: had geographers, especially those dealing with man's social geography, been too rational and positivistic in thought and practice, including applied works? Had they been too interested in developing and using mathematical, statistical and chorographical models and techniques? When condemning physical-geographical determinism, had perhaps Swedish geographers forgotten that man lives by a society of economy but in a society of ecology?

This phase also saw Swedish geographers becoming not only researchers in a more comprehensive and profound manner about and with people. They also came to work among the people and take a direct part in the daily life of man, for instance by making observations and enquiries, measuring and mapping on the spot or writing diaries etc. in order to see how a person or a family spent their time and earned their living. All these and similar contributions towards observing our changing society had started earlier, but have only come into full flower during the last ten or fifteen years.

This led to international and interdisciplinary research and a search for the roots - be it the single man or family or within the different communities and academic subjects (for Sweden see for instance Bergsten 1980 and Alvstam 1982; corresponding examples for Denmark are Biilmann 1981 and for Finland Yli-Jokipii 1982).

If we take a closer look at the research performed during this phase we find that there were some new trends in agrarian and rurally orientated geographical research towards a common understanding of human territoriality. This work has been in part quantitative e.g. Hannerbert (1976). Other work has sought the connection between land-use, laws and rules, and administrative systems, as for example Carlsson (1979), who argues that

> the development of the cultural landscape is subordinated to and affected by many factors. An important, and for the interpretation of the change of the prehistoric cultural landscape, partly neglected factor is boundaries of different kinds. Boundaries laid down between separate territories at an early stage of the development of the agrarian landscape have probably to a great extent affected and determined the continuing change of the cultural landscape.

There is also Malmberg's (1980) survey of behavioural territories in man, where he concludes:

> In a vertebrate community not governed by absolute hierarchy alone the individual acquires not only personality through his territory but also has some space where he is superior to all other members of the community. Thus, territoriality is not at all, as is sometimes claimed, a reactionary device, an antithesis to the open society. Instead, it offers perhaps the only natural and thus the only solid basis for the two cornerstones of a democracy: freedom and equality.

Many Swedish geographers have been engaged in aesthetical matters over the years, earlier concerning physical qualities, later as regards built-up areas and other items in the landscape formed by and for man. One of these studies is that of Kallsbo and Jonsson (1976) dealing with the urban-rural fringe of, among others, the towns of Malmo and Lund.

Such new ideas and works in social geography are not limited to the outer landscape or its visible face. These trends also apply to buildings and local areas. Swedish geographers have been involved in the revitalisation of old urban and industrial milieus in order to give them new leases of life or a higher degree of diversification (see e.g. Johansson 1984).

There have also been studies of man in relation to nature and the natural environment. Thus, for example, Wahlstrom (1977) argues that research and planning concerning the environment need a "united historical-ecological-geographical perspective". This holistic approach may also be exemplified by reference to recreation. Aldskogius (1981) considers the circumstances and conditions impeding the use of land for recreational purposes, even if the land in question might be perfectly suitable and attractive for outdoor recreation.

During the 1950's and 1960's urban geography was dominated by research programmes dealing with ideas and methodologies relating to national and regional systems, even if there were also studies relating to local conditions and problems. While quantitative analyses and modelling of urban systems continued in the 1970s (e.g. Andersson 1970, Nordstrom 1971, Falk 1976) localised studies became more important, for instance studies of the quality of life and living conditions and the distribution of people by groups in different places.

The last-mentioned group includes Lindberg's (1972) study of Umea and Gosta, Forsstrom's (1973) about "Malmberget", one of the mining settlements in Norbotten, founded in the late 1800's. In this study, Forsstrom gives a long-term review of the changing conditions of the mine-workers and their families. Othe reports in this group are the studies of Uppsala by Aldkogius and Aldskogius (1977), as well as Persson's (1977) study of central city employment in Gothenburg and that of Nordin (1983) on the role of the market places in i.a. Paris, not only in a commercial sense but also in people's social context.

Related to the location and use of urban elements there are studies relating to the accessibility to different kinds of service amenities. Also here the ideas and reality described lead back to the work of early geographers, such as the umland studies by Sten De Geer and Helger Nelson. What appeared during the late 1960's and in the 1970's was the use - more or less - of the time-geography approach, based on detailed data concerning peoples' possibilities and living habits - from daily business and duties to the entire life cycle - measured in time and not, as was usually the case, in money.

This was a gradual process, the first expression of which appeared in an Official Report of the Swedish Government in 1970 about "The use of time and the structure of the environment" and its changing nature.

A summary is provided by Martensson (1981) in honour of Torsten Hagerstrand on his 65th birthday. She suggests that

> In the beginning it was easy to make presentations of time geography. Listeners expected "new ideas", not ready results. Today, more than a decade later, presentations are pedagogically much more challenging. No single book covers the subject .... Much of time geography also seems self-evident .... the main objective in presenting time geography must be to arouse the curiosity of those interested, inducing them to read more and to try things out themselves.
>
> Time geography inspires individual oriented studies that are close to everyday life. Everyone can test the approach on their own situations and experiences. The chief merit of time geography then, would be that it can increase our understanding of our own circumstances and thereby increase our abilities to help ourselves. The greatest challenge in the near future is thus to find suitable means and methods of conveying these ideas to ordinary people.

Time geography added a consideration of temporal accessibility to that of spatial accessibility (e.g. Weissglas 1975, Oberg 1976). Oberg saw his work on the problems of physical access to service supply (legal aid, dentists, etc.) as an "effort to fuse traditional geographical studies of place-use with studies of time-use".

Time geography has a strong quantitative dimension, which can be illustrated by the work of Lenntorp (1976). To derive individual paths he constructed a simulation model (named PESASP, Programme Evaluating the Set of Alternative Sample Paths). The model operates with two main elements:

- activity programmes at individual level
- an environment described in space-time.

But the approach has also focused directly on the real problems of individuals with repect to work relocation, changing living conditions and distribution of child care facilities (Martensson, 1979). According to her abstract, she analysed

> The relationship between environmental structure and the growth of individual biographies. The problems considered primarily concern a biography's construction, e.g. the possibilities of fitting in different blocks of time-use and forming them into a smooth continuum. The composition of daily life is seen as a chain of activities, in various sub-environments, in interaction with others, mostly in a societal setting that involves a high degree of specialisation and a striving for sectoral effectiveness.

The time-geographic approach, she continues, thus

> stresses the interwoven character of events and situations in daily life. The elements move, merge and separate - leaving one situation to form another in the same place or elsewhere. Situations are viewed as rooted in past situations, activities as components linked together in sequences. Also, time-geography offers tools for investigating fundamental physical constraints that effect the formation of situations.

Time-geographic studies are also related to past interests, e.g. traditional regional colonisation and "frontier" studies emanating from geographers in Uppsala and Umea. Wiberg (1983) is also strongly influenced by the modern approach of accessibility of time-geography.

This expose of the studies on accessibility and time-geography in Sweden can also be used as a contribution of research on research, as i.a. J.D. Bernal and Mario Bunge have pointed out. Originating from the idea of a single man - Torsten Hagerstrand - these scientific means developed to a very high specific level.

The methodology of time-geography also vitalised the geographies of employment and of the quality of life in the light of ambitions to achieve spatial equality. Of course, such subjects had been studied for a long time (cf 1920s and 1930s) but during the

1970s and later, these matters became a very topical field of geographical research. The reasons for this increasing interest within geography can be summarised in four main points, namely:

1) Changes and financial difficulties within several fields of manufacture (primarily the textile industry but also shipbuilding and some older engineering industries) as well as within the mining of iron ore and basic iron production,
2) State or local authority subsidising of employment opportunities in existing work-sites or initiation of new such sites where old industries became redundant,
3) Increasing development within some industrial sectors (the motor industry, electronics, the chemical industry),
and
4) Rapidly increasing employment within Government administration in the capital, Stockholm, and similar tendencies in some other cities with ever growing administrative offices for county councils etc. or as locations for head offices etc. of banks and companies within expanding manufacturing or consulting sectors.

All these changes had positive or negative effects on the communities where the plants or offices were located and - most important of all - created a general feeling of uncertainty among ordinary people in places facing problems. The political aims to make Sweden a country of equality tended to become impractical in reality. Tornqvist (1970) wrote that:

> increased regional differentiatioon in population and employment has accentuated regional inequalities in income, social structure and education.

As a result of these tendencies, many scientists commenced research on employment problems. When the geographers applied themselves in strength to labour market research around 1970 these studies changed from being applied to large economic and statistical systems or aggregates to more individual and socio-geographically orientated questions. For example, Berger (1973) studied socio-psychological, medical and geographical effects of the closing down of industries in Arvika commune, Varmland county.

Later, Berger et al (1983) also investigated similar problems for ironworks and engineering industries.

Gonas (1974) examined the employment crisis at the Oskarshamn shipyard. She argued that

> The grounds for investigating this closure resides mainly in the form of support measures put in by the society state and local government. Two crises affected the shipyard, the second one leading to closure. The first crisis occurred in 1962, when the enterprise issued a warning of lay-offs for some 500 employees. The second one occurred in 1967 and affected 450 employees, most of whom were older skilled workers. To alleviate the consequences of both crises extensive labour market policy measures and municipal support were put in at a total cost of approximately 100 million Sw. kronor.

An examination of the textile industry in the context of the international division of labour was undertaken by Gustafsson (1983). This study is based on the changes, particularly during the last 30-35 years, in three Swedish textile companies with plants in Sweden and abroad, and, due to the specialisation and concentration of production in Sweden, many of the communities in which companies have been located have suffered unemployment problems. Gustafsson gives accounts of the changes and their consequences for the workers, especially the women in the smaller and isolated localities (cf Harvey 1982).

Some other work has looked at administrative and commercial functions and the changes in their localisation, e.g. Sahlberg (1970), Ahnstrom (1973), Persson (1974) and Pettersson (1980). Petterson examined the relocation of government activities from Stockholm to a number of regional centres in the 70s. The aim of the study is to measure the local labour market effects of this relocation against the background of Swedish regional policy. This is accomplished by studying the turn-over of vacant jobs in the relocated authorities in relation to vacancies in the local labour market as a whole. The basic assumptions are that people wish to remain in their home regions and that the aim of regional policy is to make this possible. Another assumption is that accessibility to work is the crucial factor in this respect. In the long term, relocation of government activities can be supposed to contribute positively to the goals of regional policy.

The time-geography approach has been used also at the very local sector (see Ellegard 1983). She suggests that

> when the individual worker performs his work-tasks, his personal experience is affected, habits are formed, his body is exposed to wear and tear, and jargons are developed. Such patterns originate in the "work culture", but may affect other activities during the rest of the individual's daily life.

There is also interest in the informal sector, where persons and families involved barter jobs and commodities without formal organisation or are working at a self-sufficiency level. For example, three current research projects at the Department of Human and Economic Geography, Gothenburg, can be mentioned: Development of the formal and the informal economy in a local community - possibilities and restrictions, (Ulla Herlitz); Social development - ethics, strategy and geographical synthesis - an example from the village of Ostmark in the northern part of the county of Varmland, (Gerhard Gustafsson); and Local development in the urban district of Kortedala, Gothenburg (Ingrid Johansson et al).

Technological changes as socio-geographical research themes can be viewed in two dimensions. The first is technological development within the discipline itself, such as mapping methods the use - perhaps also the misuse - of computers with or without digitisation equipment, etc. A discussion concerning this dimension is, however, not relevant in the present survey. The other main groups belonging to these themes cover questions of the technical effects on peoples' everyday life, viewed by geographers as part of changing society. A common review of these items was published by Lunden (1976) and a local, historical example by Sporrong and Lunden (1975). Holmgren's (1980) investigations about the effects of closing down railways can also be referred to this category.

An example from the very frontier of research is the study by Falk and Abler (1980) where they examined the effects of technology on distance, stating

> Each movement technology alters the effective distances among places by changing the effort required to move things, people, and information among those places. Intercommunications media (postal, telegraphy, telephone, telex, and computer-mediated techniques) have had greater impacts on effort distances than any other movement technology. Because they have not heretofore been examined carefully, the alterations of distance produced by intercommunications media have rarely been incorporated into geographical thinking.

But incorporation is occurring (see Forsstrom and Lorentzon, 1983). They wish to increase knowledge regarding the connection between the introduction of data communication technology and the settlement structure. The first empirical study is based on the use of "modems" in the western parts of Sweden.

While medical issues were examined before the 1970's, they then usually consisted of official reports on the occurrence of various illnesses, geographic data in the form of maps, etc. Three new lines of enquiry in more depth have since been established, viz.

1. The utilisation of geographical planning models at the local level, such as towns and rural districts, e.g. for the use of out-patient health services in the county of Vasterbotten (Weissglas, 1975).
2. The use of medical or dental service points - hospitals, surgeries, etc. - for studying the accessibility of such service to the population in different parts of the country and when using varying means of transportation, e.g. Holm (1970). Vilhelmson's (1983) study regarding travel restraints for elderly and handicapped people is a further example.
3. A form of action research to explain peoples' unhealthy environment or exposure to accidents, e.g. Forsstrom's (1982) study of commuting accidents and their background in some Swedish regions during 1971. Forsstrom (1984) finds that half of all commuting accidents are caused by environmental factors. Although environmentally caused accidents are often less severe than other accidents they do constitute more than a quarter of all days lost due to ill-

health.

Investigations regarding migration have been an important subject within Swedish social geography for a long time, as pointed out above. A new dimension has led to the use of the migration theory, data and findings within the planning and forecasting of regional and local labour markets as well as housing (see Oberg 1979, Borgegard and Haggstrom, 1981). The latter underline the causal complexity and complicated course of events as regards the flow of migration between Finland and Sweden. They conclude

> The migration steps can thus be seen as an adjustment of satisfying different kinds of needs:
>
> - the migration to Sweden usually has economic reasons. The migrant tries to satisfy basic needs and reach a better standard of living.
> - the second decision involves very much social needs - the contacts with kin and friends are important for quality of life. The migrant settles down where he has an already knitted social network.
> - from the new location the migrant makes the third adjustment, he tries to get a better dwelling. Furthermore, he improves his material standard of living.

Migration has also been linked to problems of urbanisation (see Ortenblad and Ofverholm 1983).

Studies of societal conflicts and choices for the future as viewed and solved by Swedish geographers have many intellectual, political etc. roots. Here I shall deal with only two of them. The present basic interest derives, I believe, from geographers working together with people at the regional or local level, being engaged in particularly popular research sectors or planning in close contact with the problems and intentions of ordinary people. This view is partly identical with that of other academic subjects such as political and economic science - particularly "the school of public choice" - as well as psychology, anthropology.

Regarded from a more internal point of view in geography this theme can also be seen as a revitalisation - one among many others - of the "old" geographical view of territories and boundaries and of the geopolitical thinking in the early part of the 20th century. There is, however, a very clear difference between the old geopolitical thoughts and the modern ideas. The former were seen - and used - from the macro level and the perspective of the dominating state, whereas the present ideas are viewed from and used for the benefit of the level of ordinary people or groups of people. Also here we thus witness a fruitful development from large aggregates to smaller units.

Some of the above-mentioned problems have been dealt with by Arell (1977) in Umea. He has investigated the nomadic reindeer-raising people in the border district between the northern parts of Sweden, Finland and Norway. Arell discussed the results of the political negotiations which legalised the free right of pasture for the nomadic reindeer-owner in all three countries. The conflicts between the Lapps and the new settlers have recently been studies by Stenman (1983).

One of the systematic introducers and users of this theme is also Reinhold Castensson (1979, 1980). He examines conflicts between different groups of interested parties, such as land-owners, local politicians, expanding companies, etc., with the aim of achieving the best future results for ordinary people in a countryside area. One can furthermore mention Olander's (1984) study about the sociogeographic effects of the amalgamation during the 1960's of Sweden's more than 1000 municipalities to a total number of only 400 (see also Khakee, 1983).

Swedish geographers have taken an increasingly active part, particularly during the last decades, in the "development" of what has been called "The Third World". Like John Dwyer (1977), they have often posed the question "development for whom?" - see for instance Lundqvist (1975). He showed that a person engaged in manufacturing earns at least two to three times more than a farmer and that the differences between the two categories is widening rather than narrowing. Here, the "development" or "modernisation" has increased the segregation within Tanzanian society! The unequal gains from development are also shown by Rudengren (1981), dealing with Tanzania, and Gyllstrom (1977) and Narman (1984), both about Kenya. There has also been interest in Sri Lanka (Ohrling 1977). And further, these developmental and ecological concerns have been related to other themes in Swedish social geography, especially time-geography (see Carlstein 1980 and Torell 1984).

This theme has also allowed geographers to think about the integration of academic and life practices in "terms of the future" (see Claeson, 1982, from whom the following final words are taken):

> "The integrative approach in development planning has many dimensions. One is the man-centredness. It is on the individual and domestic levels that the ultimate result of development efforts should be evaluated. One of the entries in the human dimension is the individual man. Corresponding to this approach .... are parts of human and humanistic geography, whose integrative possibilities should be exploited in area development's involvement of humans. This is valid for geographical ideas on regional identity and "perceptible interdependence". It has also a bearing on the now dominant issue of man's environment. Time-geography's combined time and space aspects on the individual's immediate environment were mentioned as an academic achievement capable of reinforcing considerations about area development.
>
> The intention of this essay has been to identify coincidences between academic geography and practical development work in the third world. Perhaps my somewhat eclectic reflections also turned into an advocacy of the presence in that part of the world of experts with an occidental upbringing and training?

Summary and Future Prospects

This study is an extract with examples from a larger work dealing with the long-range development of thoughts and practices within Swedish human geography. Within this I have thus identified five, partly overlapping, main phases since geography became an independent academic discipline about one hundred years ago. The basic subject of each relay is termed "territory", to which I have added two catchwords, both expressing the dominating ideas or themes of the respective phase.

Briefly, I have shown how social geography,

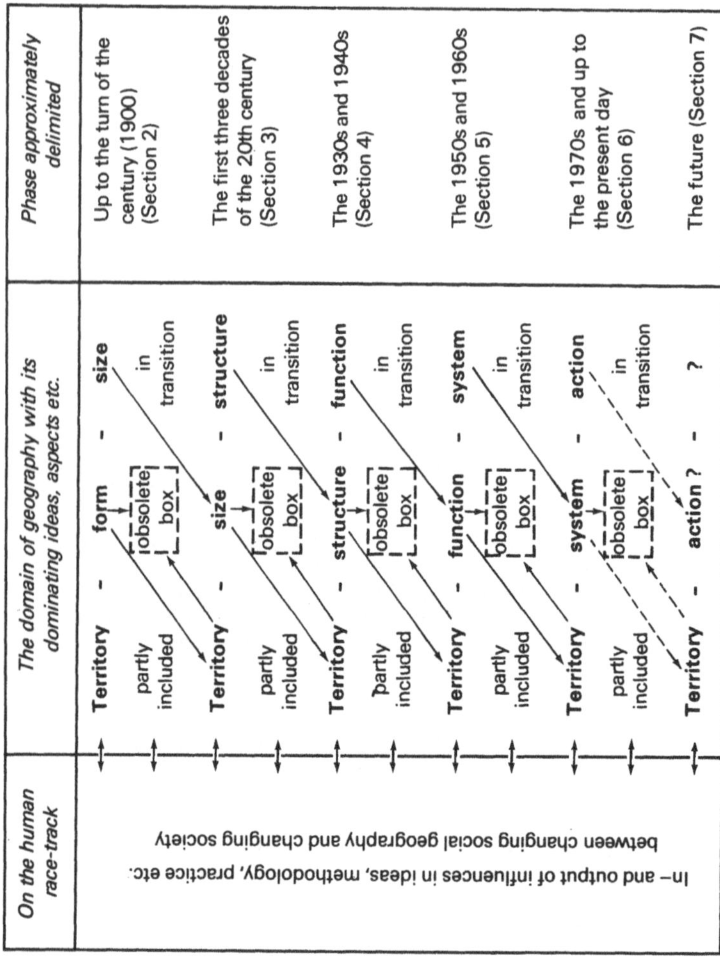

Fig. 4.1 A process chart for the study of Swedish social geography

during the first two to three decades, was dominated by anthropological, ethnographical and historical views. During the first decades of the 20th century economic aspects were added and previous interest in discussing form and size was augmented by structural ideas. Gradually, around the turn of the decade 1920/30, this phase passed more and more to doctrinal thinking about function. In the 1940's, there was a counter-reaction against physical and historical determinism. Social geography began to acquire a profile of its own. This, in turn, resulted in increasingly sophisticated and mathematically accentuated analysis of large systems. However, this quantitatively orientated methodology encountered resistance round about 1970, leading to an approach with greater emphasis on the individual and entailing more local action on the part of the researcher. The ideas and thoughts, themes and transitions expressed in this paper can be summarised in Figure 4.1?

What of the future? I believe that science is best served by a minimum of control. The account I have given above surely illustrates quite convincingly that unfettered research renews itself. However, from my point of view, with more than 40 years' experience of theoretical and applied geography and being on the verge of retirement, I should like to enumerate below some particularly interesting fields of research within social geography:

1) Geography should become involved in the concept of public choice, which to-day only comprises macroeconomics and social science. In future, there must be an increasing need of a regional or local scale, close to the common people - compare for instance with present-day politics concerning agriculture, mining etc.
2) Thinking of (1) leads to a new form of human geopolitics, far removed from Mackinder and Haushofer but close to Malmberg (1980) and Torell (1984).
3) (2) brings new thinking within human ecology - to-day ecology is mostly used as another term for biology, etc. I should like to emphasise Man in this connection. See for instance the report by Ulrich (1984).
4) The context of points 1-3 will, I believe, have a future influence on the new geolinguistics - compare with for instance the ideas of Colin Williams (1981) and others - both at the individual and international levels.

5) Societal planning for the future in a society subjected to restrictions and contractions ties in with (1). Important elements in such planning can, for instance, be data processing and telecommunications in their socio-geographical context but also, at a more human level, studies of the geography of inequality in various respects, as stressed by Smith (1979) and Alvstam (1982).

The necessity of such research and planning is aptly summarised by Nordstrom (1984),

> As a citizen, I cannot avoid reflecting upon the curious fact that, in times of economic stagnation, a diminishing energy is devoted to long-term questions while, in times of economic progress, these are considered very important. When you have plenty of money, it is not so important to care about long-term planning of the housekeeping budget. If there is a shortage of money, however, it can be valuable to carefully consider which steps to take and what the subsequent long-term consequences may be. This private knowledge apparently meets with difficulties when transferred to the public level and has scarcely influenced the debate on societal planning."

REFERENCES

Aagesen, Aa. 1949 Geografiske studier over jernbanerne i Danmark (Summary: Geographical studies on the Railways of Denmark), Kobenhavn: H. Hagerups boghandel. Also in The Royal Danish Geographical Society, communications in human Geography

Ahnstrom, L. 1973 Styrande och ledande verksamhet i Vasteuropa, en ekonomisk-geografisk studie (Summary: Governing and Administrative Activities in Western Europe),

		Stockholm: The Economic Research Institute at the Stockholm School of Economics
Ahlenius, K.	1897	Om klassifikationen af manniskoraserna ("About the classification of the human races"), Spakvetenskapliga sallskapets i Upsala forhandl. 1894-1897, 44-69
Ahlenius, K.	1906	Friedrich Ratzel och hans antropogeografiska larobyggnad ("Friedrich Ratzel and his doctrinal system in human geography"), Ymer 26, 36-57
Ajo, R.	1957	New Aspects of Geographic and Social Patterns of Net Migration Rate. A Pilot Study based on Finnish Statistics for the Year 1951, In: D Hannerberg, T. Hagerstrand, B. Odeving (eds), Migration in Sweden, A Symposium, Lund Studies in Geography B:13, Lund: C.W.K. Gleerup Publishers
Aldskogius, H.	1967	Vacation house settlement in Siljan region, Geografiska Annaler 49 B, 69-95
Aldskogius, H.	1969	Modelling the Evolution of Settlement patterns. Two Studies of Vacation House Settlement. Geografiska regionstudier 6, Uppsala: Department of Human Geography, University of Uppsala
Aldskogius, H.	1981	The Recreational Landscape Ltd, Aix-en-Provence: Centre des hautes etudes touristique, études et memoires

Aldskogius, H. & M. Aldskogius 1977 Nagra kartor over Uppsalas sociala geografi ("Some maps of the social geography of Uppsala"), In: M. Aldskogius (ed.), Uppsala. Samhalls geografiska studier, Uppsala: Department of human geography, University of Uppsala, Geografiska regionstudier 12, 94-114

Alvstam, C.G. 1982 Ekonomisk geografi ("Economic geography"), In: B. Dynefors (ed.), Knowledge and perspectives in economic sciences, Gothenburg: Centre for Interdisciplinary Studies of the Human Condition, University of Gothenburg and Chalmer's University of Technology

Ameen, L. 1964 Stadsbebyggelse och domanstruktur. Svensk stadstuveckling i relation till agoforhallanden och administrativa granser (Summary: Urban Settlement and Domain Structure. Urban Development in Sweden in Relation to Proprietary Rights and Administrative Limits), Communications from the Department of Geography, Dissertation 46, Lund: CWK Gleerup

Andersson, L. 1970 Rumsliga effekter av organisationsforandringar. Studier i lokalisering med exempel fran skolvasednet. (Summary: Spatial effects of organizational changes. Location studies with

		examples from the public education system), Gothenburg: Communications from the Department of Geography, University of Gothenburg, B:17
Andersson, L. & R. Olsson (eds)	1975	Manniska och miljo. Studier av samspelet mellan manniskans aktiviteter och utformningen av den regionala och lokala miljon (Summary: Man and his environment. Studies of the interplay between the activities of man and the shaping of our regional and local environment), Gothenburg: Gothenburg club of geographers, also Communications from the Department of Geography, University of Gothenburg B:45
Arell, N.	1977	Rennomadismen i Torne Lappmark ("The Reindeer-nomadic System in Torne Laplander's territory"), Umea: Department of Geography, University of Umea, Communication 24
Berger, S.	1973	Forstagsedlaggning - konsekvenser for individ och samhalle (Summary: Plant shut down - consequences for man and society), Uppsala: Department of Human Geography, University of Uppsala, Geografiska Regionstudier 9
Berger, S., M. Lundmark, & A. Malmberg	1983	Lokal och regional utveckling. En teoretisk ansats. Delstudie inom Bergslagsprojektet ("Local and regional development.

A theoretical attempt. Partial study within the Bergslagsproject"), Uppsala Research reports from the Department of Human Geography, University of Uppsala, 82

Bergsten, K.E. 1980 *Forskartraditioner i kulturgeografi vid svenska universitet under 1900-talet* ("Research traditions in human geography at Swedish universites during the 1900s"), Lund: Department of History, University of Lund

Biilmann, O. 1981 *Geografi, tradition og perspektiver* ("Geography, tradition and perspectives"), Brenduerup, Denmark: Geografforlaget

Bohlin, M. 1982 *The spatial and economic impact of recreational expenditures and sales in the Pigeon Lake area of Alberta*, Department of Human Geography, University of Uppsala, research report 77

Borgegard, L.E. & N. Haggstrom 1981 Migration Decisions in Theory and Practice: Finland to Sweden after 1945, In: J.W. Webb, et al (eds), *Policies of Population Redistribution*, Publicationes Instituti Geographici Universitatis Ouluensis 74, 125-140

Bunge, M. 1967 *Scientific Research II. The Search for Truth.* Studies in the Foundations, Methodology and Philosophy of Science, 3/II, Berlin, etc:

		Springer-Verlag
Buttimer, A.	1983	The practice of geography, London and New York: Longman
Bylund, E.	1956	Koloniseringen av Pite lappmark t.p.m. ar 1867 (Summary: The Colonization of Pite Lappmark until 1867), Geographica 30, Uppsala: Department of Geography, University of Uppsala
Bylund, E.	1966	Glesbygdsproblemet - Norrlandsproblemet ("The problem of sparsely populated areas - the problem of the northern part of Sweden"), Ymer, 38-54
Carlsson, D.	1979	Kulturlandskapets utvecking pa Gotland. En studie av jordbruks - och bebyggelseforandringar under jarnaldern (Summary: The Development of the Cultural Landscape on Gotland. A Study of Changes in Agriculture and Settlement during the Iron Age), Stockholm: Department of human geography, Communication B 49
Carlstein, T.	1980	Time Resources, Society and Ecology, Communications from the Department of Geography, University of Lund, Dissertations LXXXVIII
Carlstein, T. et al (eds)	1978	Human Activity and Time Geography, vol 2: The Lund School, London: Arnold
Castensson, R.	1979	Local Politicians and Local Planning. Maintenance of Knowledge and Exchange of Experiences in Local Planning, Stockholm:

		Research into Local Planning Processes, Swedish Council for Building Research D: 6
Castensson, R.	1980	*Selections for the future - on the selection of land use alternatives and the influence of local politicians on municipal planning*, Communications from the Department of Geography, University of Lund, Dissertations LXXXVI
Chabot, G.	1948	*Les Villes. Apercu de géographie humaine*, Paris: Collection Armand Colin 250, Section de Géographie
Christaller, W.	1933	*Die zentralen Orte in Suddeutschland*. Eine okonomisch-geographisce Untersuching uber die Gesetzmassigkeit der Verbreitung und Entwicklung der Siedlungen mit stadtischen Funktionen, Jena
Cleason, C.-F	1964	En korologisk publikanalys. Framstallning av demografiska gravitationsmodeller med tillampning vid omlandsbestamning pa koordinatkarta (Summary: A Regional Chorological Analysis of the Cinema Audience. Construction and Application of Demographic Gravity Models), *Geografiska Annaler* XLVI, 1-130
Claeson, C.-F	1969	A Two-stage Model of In-Migration to Urban Centres: Deductive Development of a Variant of the Gravity Hypothesis, *Geografiska Annaler* 51 B, 127-138
Claeson, C.-F	1982	Integrated development

		and integrative geography - some reflections and impulses, <u>Geografiska Annaler</u> 64B, 97-103
Dahl, S.	1955	En ny stadstyp ("A new type of town"), In : K.E. Bersten et al (eds), <u>Staden, nagra stadsbygdsproblem</u>, Gothenburg, 19-34
Dahl, S.	1967	<u>Retur till landet</u> (Summary: Back to the countryside), PLAN 3
Dahlberg, S.	1969	Regional osakerhet - en metodstudie over antal konkurser i Sverige aren 1965-1967 sasom matt pa regional instabilitet (Summary: Regional uncertainty - A methodical study of bankruptcies in Sweden during 1965-1967 as a measure of regional instability), <u>Choros</u> 17, Gothenburg: Department of Human Geography, University of Gothenburg
De Geer, S.	1919	<u>Befolkningens fordening i Sverige</u>. Beskrivning till karta i skalan 1:500 000 ("Map of the distribution of population in Sweden, Jan. 1st 1917"), Stockholm: Wahlstrom & Widestrand
De Geer, S.	1923	On the definition, methods and classification of geography, <u>Geografiska Annaler</u> V, 1-37
Dwyer, D.J.	1977	Economic Development: Development for Whom? <u>Geography</u> 62, 325-334
Eldblom, L.	1968	<u>Structure foncière, organisation et structure sociale</u>. Une étude comparative sur la vie socio-economique

dans les trois oasis libyennes de Ghat, Mourzouk et particulierement Ghadames, Lund: Communications from the Department of Geography, University of Lund, dissertations LV.

Ellegard, K. 1983 Manniska - produktion. Tidsbilder av et productionssytem (Summary: Man - produktion. Time-pictures of a production system), Gothenburg: Communications from the Department of Geography, University of Gothenburg, B:72

Enequist, G. 1943 Yrkesgruppernas fordelning i Sveriges kommuner ar 1930. En geografisk oversikt ("The Distribution of Main Occupational Groups in the Communities of Sweden, 1930"), Geographica 13, Uppsala: Department of Geography, University of Uppsala

Enequist, G. 1946 Yrkesgruppernas fordelning i Sveriges kommuner ar 1940. En geografisk oversikt (Summary: The Distribution of Main Occupational Groups in the Communities of Sweden, 1940), Geographica 18, Uppsala: Department of Geography, University of Uppsala

Falk, Th. 1976 Urban Sweden. Changes in the Distribution of Population - the 1960s in Focus, Stockholm: The Economic Research Institute at the

		Stockholm School of Economics
Falk, Th. & R. Abler	1980	Intercommunications, Distance, and Geographical Theory, _Geografiska Annaler_ 62B, 59-67
Forsstrom, G.	1973	_Malmberget, malmbrytning och bebyggelse_ ("Malmberget, mining and settlement"), Lulea: Norrbottens museum
Forsstrom, A.	1982	_Commuting Accidents_. A study of Commuting Accidents and Casualties in some Swedish Regions during 1971, Gothenburg, Communications from the Departments of Geography, University of Gothenburg, B:69
Forsstrom, A.	1984	Methodological Approach to the Study of Commuting Accidents, _Geografiska Annaler_ 66B, 59-70
Forsstrom, A. & S. Lorentzon	1984	_Datakommunikation och bebyggelsestruktur_, ("Data Communications and Settlement structure"), Department of Human and Economic Geography University of Gothenburg, Occasional Papers 1984:1
Forsstrom, A. & R. Olsson	1982	_Boende i glesbygd_ ("People in sparsely populated areas"), Stockholm: Swedish Council for Building Research R 100
Frimodig, I.	1959	_Turism i Bohuslan - en ny industri_ (Summary: Tourism in Bohuslan - a new industry, Gothenburg: Communications from the Gothenburg School of Economics, Department

		of Geography 58
Frodin, J.	1925	Siljansomradets fabodbygd (Summary: The Fabod-district around Lake Siljan), Lund: C.W.K. Gleerup, also in Publications of the New Society of Letters at Lund 5
Godlund, S.	1961	Population, Regional Hospitals, Transport Facilities and Regions. Planning the Location of Regional Hospitals in Sweden, Lund studies in geography B:21
Godlund, S.	1984	Mansklig miljo i centrum ("Focusing on the human environment"), Gothenburg and Linkoping: National Road & Traffic Research Institute and University of Gothenburg
Godlund, S. & K. Gotlund	1976	Stad och omvarld. Norrkopings ekonomiska och sociala historia 1915-1970 ("City and its Surrounding World. The economic and social history of the town of Norrkoping 1915-1970"), Norrkoping: Municipality council
Goldfield, D.R.	1981	Svensk stad and American Urban History. In: J. Hammarstrom and Th. Hall (eds), Perspektiv pa Svensk stad, Stockholm: Akademilitteratur
Gonas, L.	1974	Foretagsnedlaggning och arbetsmarknadspolitik. En studie av sysselsattningskriserna vid Oskarshamns varv (Summary: Plant Shut-Down and Labour Market Policy. A Study of the Employment

Crises at Oskarshamn's Shipyard), Uppsala: Department of Human Geography, University of Uppsala, Geografiska Regionstudier 10

Goodman, N. 1978 Ways of Wordmaking, Hassocks: Harvester Press

Gustafsson, K. 1983 Tekoindustrin och forandringarna i den internationella arbetsfordelningen - konsekvenser for lokalisering och systellsattning i Sverige (Summary: The Textile and Clothing Industry and the Changing International Division of Labour - Consequences of Location and Employment in Sweden), Gothenburg: Communications from the Department of Geography, University of Gothenburg, B:73

Gould, P. 1981 Letting the data speak for themselves, Ann. of the Association of American Geographers 71, 166-176

Gyllstrom, B. 1977 The Organisation of Production as Space-Modelling Mechanism in Under-development Countries - the Case of Tea Production in Kenya, Lund: Communications from the Department of Geography, University of Lund, Dissertations LXXIX

Hannerberg, D. 1976 Models of medieval and pre-medieval territorial organisation, with a select bibliography, Journal of Historical Geography

2, 21-34

Hansen, J.C. 1983 Regional Policy in an Oil Economy: the Case of Norway, Geoforum 14, 353-361

Harvey, D. 1982 The Limits to Capital, Oxford: Basil Blackwell

Heidegger, M. 1968 Existence and Being, Chicago Ill.: Henry Regnery

Helmfrid, S. 1978 Arealkonflikter vid tatortstillvaxt - problem for en samordnad regional och fysisk planering ("Space conflicts in urban development - problems for a co-ordinated regional and physical planning"), In: Att forma regional framtid ... A report from the Expert group for regional research, ERU, Stockholm: Publica, Liber forlag

Holm, S. 1970 Accessibility to National Dental Service in the Province of Norrbotten in 1950 and 1965, Geografiska Annaler 52B, 141-152

Hagerstrand, T. 1947 En landsbygdsbefolknings flyttnings - roresler. Studier over migrationen pa grundval av Asby sockens flyttningslangder 1840-1944 (Summary: The movements of a rural population. Migration studies on the basis of the parish registers of Asby 1840-1944), The Swedish Geographical Yearbook 23, 114-142

Hagerstrand, T. 1949 Flyttningarna till och fran Simrishanm under 1900-talet ("In-and out

		migration concerning the town of Simrishamn during the 1900's"), IN: H. Nelson (ed), Simrishamn with umland, 74-111, Lund: Gleerupska Universitetsbokhandeln
Hagerstrand, T.	1953	Innovationsforloppet ur korologisk synpunkt ("Innovation diffusion as a spatial process"), Communications from the Department of Geography, University of Lund, Dissertations, XXV: Gleerupska Univ. bok-handeln
Hagerstrand, T.	1957	Migration and Area. Survey of a Sample of Swedish Migration fields and Hypothetical Considerations on Their Genesis, In: D. Hannerberg, T. Hagerstrand, B. Odeving (eds), Migration in Sweden. A Symposium, Lund Studies in Geography B:13, 27-158, Lund: C.W.K. Gleerup Publishers
Hagerstrand, T.	1967a	Innovation diffusion as a spatial process, translated by A. Pred, Chicago: Chicago University Press
Hagerstrand, T.	1967b	The domain of human geography, In: R.J. Chorley (ed), Directions in geography, London: Methuen, 67-87
Hagerstrand, T.	1982	Diorama, Path and Project. Tijdschrift voor economische en sociale geografie, 73, 232-339
Jakobsson, A.	1969	Omflyttningen i Sverige 1950-1960. Komparativa studier av migrations-

falt, flyttningsavstand och mobilitet, (Summary: Internal migration in Sweden 1950-1960. Comparative studies of migration fields, migration distance and mobility), Stockholm: Monographs in connection with the 1960 census of population and housing in Sweden, published by the national central bureau of statistics and the national housing board 5, also: Communications from the department of geography, University of Lund, dissertations LIX

Johansson, I. 1984 *Arbetsplatslokalisering i staden: Datid-Nutid-Framtid* (Summary: The Localization of Places of Work in the Town: Past-Present and Future), Gothenburg: Communications from the Department of Geography, University of Gothenburg, B:74

Johnston, R.J. *Geography and Geographers. Anglo-American Human Geography since 1945*, London: Arnold

Jonasson, O. 1942 Svenskarna i varlden (Summary: The Swedish Element in World-population), *The Swedish Geographical Yearbook* 18, 395-409

Jonasson, O. et al 1957 *Industriutveckling och industrilokalisering i Goteborg i jamforlse med Stockholm, Malmo och Riket* ("Development and Localization of the Industries in Goteborg in comparison with the

		Industries in Stockholm, Malmo, and Sweden as a whole.") Gothenburg: municipal council.
Jones, E. & J. Eyles	1977	An Introduction to Social Geography, Oxford: University Press
Jarbur, G.	1964	Befolkningsstrukurens forandring i Biskopsgarden, Goteborg. En social-geografisk undersokning (Summary: Structural Changes of the Population in Biskopsgarden, Gothenburg - A Social Geographical Study), The Swedish Geographical Yearbook 40, 107-115
Jarhult, D.	1958	Arbetskraftrekytoringen vid Norrbottens Jarnverks AB. En migrationsundersokning (Summary: The Recruiting of Labour at Norrbottens Jarneverks AB. A migration investigation), Stockholm: Department of Geography
Kant, E.	1926	Tartu, Etude d'un environment et organisme urbaine, Tartu
Kant, E.	1948	Den sociologiska regionen, den sociala tiden och det sociala rummet (Summary: About sociological regionalism, social time and social space), The Swedish Geographical Yearbook 24, 109-132
Khakee, A.	1983	Municipal planning. Restrictions, methods and organisational

problems, Umea: Department of Geography, University of Umea, Communication 27, also in Swedish Council for Building Research, D14

Kuhn, Th.S. 1962 The Structure of Scientific Revolutions, Chicago, Ill.: University of Chicago Press. 2nd edition 1970

Kallsbo, A. & I. Jonsson 1976 Regional miljo ("Regional milieu"), Malmo: The association of the municipalities in southwestern Scania.

Lenntorp, B. 1976 Paths in Space-time Environments. A time-geographic Study of Movement Possibilities of Individuals, Lund: Communications from the Department of Geography, University of Lund, Dissertations LXXVII

Ljungdahl, S.G. 1938 Sommar-Stockholm ("Summer-Stockholm"), Ymer 58, 218-242

Lewan, N. 1967 Lansbebyggelse i forvandling. En studie av utvecklingen i Skane sedan 1910 med sarskild hansyn till arbetstillgallenas omfordelning (Summary: Rural Settlement in Transition. A Study with Special Reference to the Concentration of Employment Opportunities), Communications, from the Department of Geography, University of Lund, Dissertations L, Lund C.W.K. Gleerup

Lindberg, O. 1972 Urban settlement and income distribution. A

		geographic study of the city of Umea, Umea: The Department of Geography, University of Umea, Geographical Reports 3
Lunden, Th.	1973	Individens rumsliga beteende i ett gransomrade (Summary: Individual spatial behaviour in a boundary area), Stockholm: Department of human geography, University of Stockholm, Communication B26
Lunden, Th.	1976	Teknik och samhalle ("Technology and Society"), Stockholm: The Committee for Future Oriented Research
Lundqvist, J.	1975	Local and central impulses for change and development. A case study of Morogoro District, Tanzania, Ad Novas - Norwegian Geographical Studies 12, Bergen: Universitetsforlaget, also in Communications from the Departments of Geography, Univeroity of Gothenburg, B:44
Lysen, A.	1955	Intensitet och distans, Studier av telefontrfiken i en vastsvensk bygd (Summary: Intensity and Distance. A Study of the Telephone Traffic in a part of Western Sweden), Gothenburg: Gothia 7, 41-57
Lonborg, S.	1902	Finnmarkerna i mellersta Skandinavien ("The Finnish-settled regions in the middle of Scandinavia"), Ymer

		22, 65-90, 361-408, 465-504
Lovgren, E.	1957	Mutual Relations between Migration Fields: A Circulation Analysis, In: D. Hannerberg, T. Hagerstrand, B. Odeving (eds), Migration in Sweden, A Symposium Lund Studies in Geography B:13, 159-169, Lund: C.W.K. Gleerup Publishers
Malmberg, T.	1980	Human territoriality. Survey of behavioural territories in many with preliminary analysis and discussion of meaning, The Hague: Mouton. Also in New Babylon, Studies in the Social Sciences 33
Martensson, S.	1979	On the Formation of Biographies in Space-Time Environments, Lund: Communications from the Department of Geography, University of Lund, Dissertations LXXXIV
Martensson, S.	1981	Femton ars erfarenhet av att presentera tidsgeografi (Abstract: Fifteen years of experience in presenting the time-geographic approach), The Swedish Geographical Yearbook 57, 130-140
Nelson, H.	1913	En bergslagsbygd. En historisk-geografisk ofverblick ("A mining district. A study in historical geography"), Ymer 33, 278-352
Nelson, H.	1943	The Swedes and the Swedish Settlements in North America, I text

		and II atlas, Lund: C.W.K. Gleerup and Oxford: Oxford University Press, also in Acta Reg. Societas Humaniorum Litterarum Lundensis XXXVII
Nordin, Ch.	1973	Marches, Commercants, Clientelle. Le commerce non sedentaire de la région parisienne. Etude de geographie humaine, Gothenburg: Communications from the Department of Geography, University of Gothenburg, B:71
Nordstrom, L.	1971	Rumsliga forandringar och ekonomisk utveckling (Summary: Spatial changes and economic development), Gothenburg: Communications from the Department of Geography, University of Gothenburg, B:23
Nordstrom, L.	1984	Is Long-term Societal Planning Withering Away? Scandinavian Housing and Planning Research
Nordstrom, O	1957	Befolkningoutbeckling och arbetskraftsproblem i ostra Smaland 1800-1955 ("Population changes and employment problems in eastern Smaland 1800-1955"), Communications from the Department of Geography, University of Lund XXXIII, Lund: Gleerupska Univ.-bok-handeln
Nordstrom, O. & S. Martensson	1966	Turism pa Oland ("Tourism on Oland"), Lund: Communications from the Department of Geography, University

of Lund, Dissertation XLVIII

Narman, A. 1984 Regional distribution of educational facilities in Kenya, with special reference to secondary school teachers, Gothenburg: Department of Human and Economic Geography, _Choros_ 1984: 2

Oberg, S. 1976 _Methods of Describing Physical Access to Supply Points_, Lund: Communications from the Department of Geography, University of Lund, Dissertations LXXVl

Oberg, S. 1979 Regional policy and interregional migration - matching jobs and individuals on local labour markets, _Regional Studies_ 13

Ohrling, S. 1977 _Rural Change and Spatial Reorganisation in Sri Lanka_. Barriers against development of traditional Sinhalese local communities, London: Curzon Press. also Scandianvian Institute of Asian Studies Monograph Series 34

Ortenblad, O. & Ch. Ofverholm 1983 Snaering i storstadscentra. Vem vill bo centralt? (Summary: Renewal of city centres. Who wants to live in city centres?), _Danish Geographical Journal_ 83, 65-73

Olsson, G. 1965 Distance and Human Interaction. A Migration Study, _Geografiska Annaler_ 47 B, 3-43

Olsson, G. 1980 _Birds in Egg - Eggs in_

		Bird, London: Pion Limited, 2nd Edition
Paulsson, G.	1950-1953	Svensk Stad ("Swedish Town"), 1-3, Stockholm: Bonniers
Peet, R. (ed)	1977	Radical Geography. Alternative viewpoints on Contemporary Social Issues, London: Methuen
Persson, Ch.	1974	Kontaktarbete och framtida lokaliser-ignsforandringar. Modellstudier med tillampning pa statlig forvaltning ("Contact work and future changes of localization. Model studies with application to state government"), Lund: Communications from the Department of Geography, Universitry of Lund, Dissertations LXXI
Persson, K.	1977	Syssellsattningen i centrum. Sysellsattningsforandringar i stadscentrum, deras orsaker och verkan - med exempel fran Coteborg (Summary: Employment opportunities in city centres. Cause and effect of changes in employment as regards city centres, exemplified by the situation in Gothenburg), Gothenburg: communications from the Department of Geography, University of Gothenburg, B:60
Pettersson, R.	1980	Omlokalisering av statlig verksamhet. Effekter pa

argetsmarknaden i mottagande orter, (Summary: Relocation of Government Activities. Effects on the Labour market in Receiving Places), Gothenburg: Communications from the Department of Geography, University of Gothenburg, B:66, also in Swedish Government's Official Reports 1980:6, appendix 8

Pred, A.(ed) 1981 Space and time in geography. Essays dedicated to Torsten Hagerstrand, Lund studies in Geography B: 48

Palsson, E. 1958 Gymnasiers rekrytering och localisering. Samhallsgeografiska studier (Summary: The Recruitment and Localization of Swedish Gymnasiums. A Study in Social Geography), Communications from the Department of Geography, University of Lund, XXXV, Lund: Gleerupska Univ.-bokhandeln

Ratzel, F. 1982 Anthropogeographie. Grundzuge der Anwendung der Erdkunde auf die Geschichte, Munchen

Rasmussen, T.F. 1966 Storbyutvikling og arbeidsreiser. En undersokelse av pendling, befolkningsutbikling, naeringsliv og urbanisering i Oslooradet (Summary: Metropolitan growth, commuting and urbanization in the Oslo area), Oslo:

		Samfunnsokonomiske studier 18
Rudberg, S.	1957	Odemarkerna och den perifera bebyggelsen i inre Nordsverige. En diskussion av vissa orsakssamband bekom fordelningen bygd-obydg. (Summary: Unsettled Areas and Frontier Settlement Areas in Inner Northern Sweden - A Discussion of Some Causes Underlying this Division), _Geographica_ 33, Uppsala: Department of Geography, University of Uppsala
Rudengren, J.	1981	_Peasants by Preference? Socio-Economic and Environmental Aspects of Rural Development in Tanzania_, Stockholm: The Economic Research Institute, Stockholm School of Economics
Ruin, H.	1935	_Poesins mystik_ ("The Mysticism of Poetry"), Lund: Gleerups forlag, 2nd edition 1970
Sahlberg, B.	1970	_Interregionala kontaktmonster_. Personkontakter inom svenskt naringsliv - en flypassagerarstudie ("Interregional Contact Patterns - A Study of Business Travel via Swedish Domestic Airlines"), Lund: Communications from the Department of Geography, University of Lund, Dissertations LXIII
Sandler, R.	1911	_Samhallet sadant det ar_ ("Society in reality"), 2nd edition 1924, Stockholm: Tidens forlag

Schwerin, H.H. von	1891	Slafveri och Slafhandel i Afrika ("Slavery and Slave Trade in Africa"), Lund: Collin & Sickerman
Schwerin, H.H. von	1892	Muhammedianismen i Afrika ("Mohammedanism in Africa"), Lund.
Skidelsky, R.	1983	John Maynard Keynes - Hopes Betrayed, London: Macmillan
Smith, D.M.	1979	Where the Grass is Greener: Living in an Unequal World, Harmondsworth: Penguin Books
Sporrong, U. & Th. Lunden	1975	Engelsberg Ironworks - estate and community in the 20th century. A geographical account. In: B. Holtze et al, Swedish Industrial Archaeology, Engelsberg Ironworks, A Pilot Project, Stockholm: The Swedish Royal Academy of Letters, History and Antiquities, 247-363
Stenman, L.	1983	Avvittringen i Vasterbottens lans lappmarker ("Delimitation in the Lapland Region of the County of Vasterbotten"), Uppsala: Department of Human Geography, University of Uppsala, Research Report 83
Sund, T.	1947	Bergens byomrade og dets geografiske utvikling 1900-1940 ("The built-up area of Bergen and its geographical development 1900-1940"), Bergen: J. D. Byer A.S.
Soderlund, A.	1923	Befolkningens fordeling i Norge ("The

		distribution of population in Norway"), Kristiania: Norges geografiske Opmaling
Thunen, J.H. von	1826	Der isolierte Staat in Beziehung auf Landwirtschaft and Nationalokonomie, Hamburg
Torell, M.	1984	Fisheries in Thailand. Geographical studies about the utlisation of resources in semi-enclosed seas, Communicatons from the Department of Geography, University of Gothenburg, B;75
Tuominen, O.	1949	Das Einflussgebiet der Stadt Turku im System der Einflussgebiete SW-Finnlands, Fennia 71: 5, Helsinki: Department of Geography, Unversity of Helsinki
Tornebohm, H.	1975	Inquiring Systems and Paradigms. Gothenburg: Report from the Department of Theory of Science, University of Gothenburg
Tornqvist, G.	1967	TV-agandets utveckling i Sverige 1956-65, (Summary: Growth of TV Ownership in Sweden, 1956-1965. An Empirical-theoretical Study), Stockholm: The Industrial Institute for Economic and Social Research
Tornqvist, G.	1970	Contact Systems and Regional Development, Lund Studies in Geography, B:35
Tornqvist, G.	1983	Kreativitetens geografi (Abstract: the Geography of Creativity), The Swedish Geographic

		Yearbook 59, 104-123
Ulrich, R.S.	1984	View Through a Window May Influence Recovery from Surgery, Science 224, April 27, 420-421
Wahlstrom, L.	1977	Naturvarden i regional och lokal planering (Summary: Nature conservancy in regional and local planning), Gothenburg: Communications from the Department of Geography, University of Gothenburg B:59
Wahlstrom, L.	1984	Geografiutveckling och geografisk utveckling - some om platser betydde nagot (Summary: Development of geography and geographical development if places mattered), Gothenburg; Department of Human and Economic Geography, University of Gothenburg, B:76
Wallin, E.	1980	Vardagslivets generativa grammatik - vid gransen mellan natur och kultur (Summary: The Generative Grammar of Everyday Life - at the borderline between Nature and Culture), Communications from the Department of Geography, University of Lund, Dissertations LXXXV
Weissglas, G.	1975	Studies on Service Problems in the Sparsely Population Areas in Northern Sweden. Umea: Department of Geography, University of Umea, Geographical Reports, 5
Wentura, F.	1978	Kulturgeografins

		framtidsperspektiv - behov av social medvetenhet (Abstract: Future prospects of human geography - the need of social consciousnes), The <u>Swedish Geographic Yearbook</u> 45, 81-95
Wiberg, U.	1983	Service i glesbygd - trender och plane ringsmojligheter (Summary: Services in rural areas - trends and planning possibilities), Umea: Department of Geography, University of Umea, <u>Gerum</u> B:8
Wilhelmson, B.	1983	Transportsystem utformade med hansyn till aldre och handikappade ("Transport Systems formed according to the needs of elderly and handicapped persons"), Gothenburg: Department of Human and Economic Geography, University of Gothenburg. <u>Occasional Papers</u> 1983:1
William-Olsson, W.	1937	<u>Huvuddragen av Stockholms geografiska utveckling</u> 1850-1930 ("The main features of geographical development of Stockholm 1850-1930"), Stockholm: Municipality council
William-Olsson, W.	1941	<u>Stockholms framtida utveckling</u> ("The future development of Stockholm"), Stockholm: Municipality council
Williams, C.H.	1981	The Territorial Dimension in Language Planning: An Evaluation of its

		Potential in Contemporary Wales, Language Problems and Language Planning 5, 57-73
Warneryd, O.	1968	Interdependence in Urban Systems, Gothenburg: Communications from the Department of Geography, B:1
Warneryd, O.	1981	Att se det stora i det lilla (Abstract: To see the universe in the locality), The Swedish Geographic Yearbook 57, 246-254
Yli-Jokipii, P.	1982	Trends in Finnish Geography in 1920-1979 in the Light of the Journals of the Period. Fennia 160, 95-193, Helsinki: Department of Geography, University of Helsinki

Acknowledgements

The following persons have helped me to produce this paper: Hans Aldskogius (Uppsala), Lennart Ameen (Lund), Birgitta Andersson (Stockholm), Lennart Andersson (Karlstad), Erik Bylund (Umea), Reinhold Castensson (Linkoping), Claes-Fredrick Claeson (Stockholm), Arthur Green (Gothenburg), Lars Nordstrom (Gothenburg), Gosta Weissglas (Umea) and Olof Warneryd (Lund). To these collaborators and to all other persons within the field of geographical research, I wish to express my sincere gratitude. To those deceased, I wish to add my great respect for their valuable contributions to later generations of geographers and thereby also to the present day standards which Swedish social geography has gradually attained. I sincerely hope that I have not misinterpreted their thoughts and results.

Grants for the main study, of which this paper represents an abstact, have been obtained from the Magnus Bergvall Foundation and the Gothenberg Business School Foundation.

Chapter Five

SOCIAL GEOGRAPHICAL RESEARCH IN GERMANY

E. Thomale

Compared with other West European countries - notably France, Great Britain and the Netherlands - social geography in German-speaking lands can look back over only a comparatively short period of about thirty years. It began late. There was little joint study between geography and sociology, and hence little exchange of views and materials to provide a relevant stimulus to geography. Further, a wide section of German sociology for decades either worked in a speculative and theoretical rather than empirical field, or lost itself in ideologically-coloured total analyses of societies. Such theoretical work had little impact on geography and sociological studies were of practical value to geographical research only when built upon micro-social investigations, as for example studies of sociological units such as selected villages or towns. This road was early followed in the Netherlands, without initially any parallel development taking place in Germany (Vooys, 1950). Similar bridges were formed between geography and history in France, and between geography and social research in English-speaking lands (Claval, 1964; Taylor, 1953)

Points of contact between sociology and geography could have resulted from a common research objective approached from either side, such as population. Because of the absence in sociology of an empirical approach and therefore the under-development of quantitative methods in German sociology, (and a strong presence of this tradition in geography) no significant contributions were made to demography, with only a few exceptions - these remain exceptions to this day (Mackenroth, 1953).

In any case there had never been in German geography - since its beginnings - any inclination to make the social complex a research objective, or even to sample its thematic possibilities. German geography saw itself at first as physical geography, focussing upon the physiognomy of the earth's surface; and later upon the describable content and form of complex Landschaften, having reference to the settlement, economics and transport components of the Landschaft, and their spatial interrelationships. Most successful were the dominant emphases on measurable and mappable phenomena. Such analyses were dependent on sense perception, and were thus physiognomic. Occasionally there were glimpses of social-geographical thinking, particularly in relation to economic geography (Ruhl, 1938): but the absence of any significant German text on population geography since that of Ratzel can be attributed to this retarded investigation of themes in social and population studies.

Important also to understanding the late development of social geography in Germany are political circumstances, particularly during the epoch of national socialism. For science the establishment of the absolute state meant not only the end of free intellectual activity, but also - and with more significant consequences - an internal narrowing to neutral themes. A social geography at that time was neither politically nor ideologically feasible. Such geography of social groups apparently logically led to geography of social conflict, making it impossible in the existing power system. And even personal courage was useless in view of the absence of access to statistical sources, and the lack of a free press.

As soon as the sketched political and methodological pitfalls were passed, German geography was then in the position to link up with the social geography of foreign countries, to establish itself and then discover new avenues. The rediscovery of former studies, methodological borrowing and native creativity led to a fascinating German-oriented mixture.

Initial Situation

The amassing of at first theoretical and then, subsequently, empirical research studies, particularly during the fifties, revealed a constellation of distinctive lines of advance. The sudden research drive, accelerated by the arrival of a younger generation of researchers, closed the inherited research lacunae, and widened the methodological base. These features also led to a change in teaching aims in school and in university. The resumption of geographical teaching and research undertakings in the first post-war years had led to widely-voiced self-criticism of the remaining gaps in German geography. New perspectives were opened up for the subject. Possible extensions to the existing geographical teaching structure - extensions capable of development corresponding to the expectations of the increasingly-critical scientific community - were revealed in subject-areas from which German human geography with its morphogenetic research into the cultural Landschaft had separated itself; that is to say in the subject areas of population and of community. Both these forms of spatial expressions of human existence were taken together, since the empirical material and a suitable conceptual apparatus were not available to permit separate treatment. Because of the importance of man's material needs, the absorption of these needs by the population complex became an important avenue of advance in social geography, without the division of the two groupings, population and community, being at first observed. Further, the developing emancipation from a hitherto restricted range of evidence considered relevant - that had particularly hindered the development of settlement geography - took place with some indirect assistance. The search for ordering principles in the cultural landscape, a reaction against simple descriptive schemes, had begun long before. But central place research, and associated functional aspects, introduced such an ordering system, already recognised in the thirties by Christaller. This now had to be thought through from an anthropocentric viewpoint. These functional regions, particularly their activating mechanisms, could only be fully understood, if the actual individual actors were added. Research, therefore, commenced to correlate the social division of labour with the functional differentiation of space, interpreting at the same time the disposition of social needs. The widespread hope of discovering the sources of spatial-ordering forces within the social region, and thus the social genesis of territorial organisation, stimulated yet further social geography (Hottes, 1968; Storkebaum, 1969).

And then the regional geographers - with their still unquestioned aim of studying the spatial interrelationships of physical and human geography - developed an interest in this new avenue of approach to the social region. Demanded of the regional geographers aware of these advances was a social geography able to provide a key to regionalisation based not on idiography, but on the analysis of social structures; to provide a new methodology for examination of the material of regional geography. As an example of this development in regional geography, to render intelligible the order of presentation of material in regional texts, we may cite the attempt to treat regional themes from the viewpoint of the spatial needs of a community (Hartke, 1963; Sievers, 1964).

At the same time academics in geography could not ignore the changes in orientation in school geography, stemming in the fifties from a demand for more actuality, more social reality and more political sensitivity in education. A demand for geographical themes that touched immediately upon the experience of scholars also increased. Indeed up to the present day, didactics have been a permanent impulse for change in geography; often a progressive approach to both didactics and geography is represented by the same person (Geipel, 1960; 1969).

Representatives of the subject taking such a broad view were also aware that to provide young geographers with professional alternatives it was necessary to increase the social sciences content and methodology within geographical teaching. Geographers could then compete with economists and other social scientists in, for example, the planning field.

Establishment

Without further initiatives and methodological leads these needs and wishes could not be expressed in research directives. The necessary measurements and the first development of themes were not, however, long in coming. They had started by the end of the forties and - with few exceptions - were given authority first by Bobek (Vienna) and later by Hartke and Geipel (Frankfurt, then Munich). Including their students this south German innovation axis henceforth contributed in both personnel and research the greater part of German social geography, a concentration not always expeditious to the developing discipline (Gerling, 1965; Otremba, 1962; Wirth, 1969).

Despite the new openings for development, social geography was only partly within geography as it was then conceived, as not all social phenomena were considered research material. Those social elements that appeared macrogeographical - adapted to a global scale - were omitted as were those social elements considered microgeographical. What were sought were those elements of a community with spatial juxtapositions and social stratification accessible to the scale of regional geography. What were chosen were those middle-scale social elements, between on one hand the family, on the other hand highly complex social systems, whose territorial association corresponded to the scale of contemporary regional geography. Reference was made to the French concept genres de vie, and attempts were made to use such fundamental social categories as farmer, herdsman, fisherman, industrial worker - with their repetitive communal life styles in space and in community - in basic social geographical classification (Bobek, 1948).

Basically important was not a temporary (and soon anachronistic) answer but a response to the complex question on to the nature, development and real context of those elements of the social corpus of geographical relevance. The problem of the social group, which was at the same time a geographical group, and which possessed at least a minimum level of distinctiveness, stability and compactness - demanded for methodological reasons by the emerging empirical work, required in answer a permanent synchronisation of social and spatial scales so that the lebensraum of a society could be effectively distinguished. This then was the first research demand of social geography - the group concept.

A further lead was provided by Bobek's suggestion to direct the search for social spatially-effective ordering principles - in so far as these were divisible into economic and technical laws - into the search for pattern in the continuous competition between social groups for disposable space. This also permitted the employment of a working hypothesis that in this territorial competition numerous social reactions contributed to the spatial provision, as in the cases of increased population density, depopulation, nucleation and dispersal, distribution and nucleus formation. These reciprocal effects were preferably studied by means of social-spatial analyses of basic social functions. These functions included continuance (of the group), work, the exercise of power, arrangement of settlement and migration. It was recognised that these concepts were not new, but derived from the Chicago school of sociology twenty years before. On account of this origin the second theme in social geography should be called the social-ecology contribution (Bobek, 1948; Hawley, 1962).

The third theme, also suggested by Bobek in 1948, as an alternative methodology, was based on the acceptance of society as a regional phenomenon sui generis, similar to the material substrate of populations. Then geographical interpretation should regard society "als regionale Erscheinung an sich, vor allem in Hinblick auf ihre raumliche Anordnung, im Zusammenhang mit Struktur und Okologie" (i.e. as a regional phenomenon in itself, above all with reference to its spatial ordering in relation to structure and ecology, Bobek, 1948, 123), which entailed the study of localisation, aggregation and arrangement. The important "an sich" (in itself) signals none other than geographical consideration of social space, here termed social spatial analysis.

Possibly we are still in the euphoria associated with these means of explanation and analysis - if the numerous gaps in the contributions originating with Bobek can be overlooked. This comment applies particularly to the lack of any increased genetic profundity, and also to the continuing fiction that societies as a whole, or in their elements, can equally be studied as unchanging units in the distributional pattern. It must also be remarked that from the beginning there has been no evaluation of what processes at present operating in the social region were to be investigated in geography. Completely unarticulated was the question whether the wide ranging goals enunciated in theory were attainable in practice. Some themes led over the boundaries of the discipline into the area of sociology, introducing to the then geographical generation material for which they were unprepared.

It gradually came about, now unhesitatingly under the social geography badge, that a former research theme, the cultural landscape - was rejuvenated, despite its apparent, irremediable methological stagnation. Many aspects of social change are effectively imprinted in the cultural landscape and can be recognised from place-to-place, and from phase to phase. Small scale land use and social mapping down to the level of individual landparcels and individual households had shown that even short period changes in spatial patterns are linked with movements relating to the whole of a society. The changes are group specific and, in operation, represent social regularities. It is possible to suggest - with Hartke (1959) - that the cultural landscape can be seen as a registration plate of social change on which is recorded group specific alterations in social values, and from which the social processes are in turn released. The analysis of change of the spatial pattern in the landscape for social diagnosis developed strongly and was an extraordinary spur to methodology. The ease with which it was accepted reflected the shrinkage in the more traditional landscape studies. The indicator method - as it was called - is the fourth avenue of advance explored by German social geographers - and is indeed a specific German variant (Hartke, 1953, 1958, 1959).

With wider use of the indicator method it became clear the small scale analyses are ambiguous, revealing social change only at particular places and in specific situations, presenting a restricted perspective of a whole society. Research revealed complex temporal migrations between social and spatial change, and that the same time revealed that nearly all social phenomena are in continuous change. The social-spatial relationship depicted at any given moment was no more than a transitional stage of earlier-initiated processes (Schaffer, 1968a; 1968b). The social motors of such processes, the functional fields of life in social groups, dealing with housing and work, provision and consumption, education and recreation were therefore of great importance, and should not be overlooked in social geography. This led to a renewed inspection of the problem, widened to include social psychological relationships, group-typical reaction schemes and mental processes. As a method of geographical problem analysis this constituted the fifth research avenue developed by social geographers (Rupert & Schaffer, 1959).

With this reference to process analysis as the most recent framework adopted by German social geography, this presentation of the basic adopted methodologies can be brought to a close. It is to be remembered that the first generation of this movement, stimulated by the group concept, by social ecology and by social space research, owed these stimuli to conceptual imports introduced by Bobek. They were received by a unprepared body of geographers, and were still to be tested empirically. In contrast, the later-introduced indicator method - due to Hartke - and process analysis - could build on the experiences gained in the interval. These later themes or approaches spread therefore over a existing social geographical substrate, a consensus that welcomed the spirit of the new themes. Nevertheless if these later themes were won more or less by fieldwork, such fieldwork was very uneven. This is evident on inspection, which reveals clear distinctions in the degree to which the research can be applied on the intellectual basis and by the fruitfulness of the methodologies.

Research Contributions

Group Concept

As mentioned above, the beginning of social geography was linked with the concept of the social group derived from life-form group analysis. This group, of all the endless stream of social forms resulting from both natural and social environmental conditions, came nearest to geographical needs, that is for a unit to be employed in regionalisation at both the global and the regional level. It was soon clear that with the help of this concept only a coarse register had been achieved. Fine social differences, brought to light by intercultural comparison, went unnoticed. This applied particularly to mobile social forms especially in highly-developed industrial societies, to which the concept was hardly applicable. It is also possible to say that the geographical account of life-form groups was not sociological.

Up to the present there have been in German social geography only two attempts to interpret this type of social group spatially. Totten (1959) sought to distinguish Saudi Arabian examples of primary and secondary life-form groups, where the environmental contrasts between oasis and desert were reflected in the social milieu contrasts between clan peasant farming and family nomadism, and were overlain by the bureaucratic apparatus of the oil industry. The wish to achieve a geographical ordering of the social corpus was obviously only achieved at the price of employing a coarse scheme in that the social spatial pattern in reality lacked sharp division and homogeneity. A similar methodology was employed by Vogel (1959) in a contrast study of social industrial geography, using the industrial life-forms of the hard coal and the brown coal industries. According to these observations the permanent mining shafts of the Ruhr hard coal industry were associated with a qualified, permanent, settled workforce; the brown coal pits of the Lower Rhinelands were associated with a less qualified, less permanent and less settled workforce, reflecting the more temporary nature of the pits. Unfortunately these phenomena were not linked with social historical material, and the original questions not followed further.

Echoes of this concept still occur occasionally, as in the recognised contrast between alpine mountain and alpine valley communities. The concept remains little changed from its early form. But with the contemporary development of statistical methods, the abandonment of field studies, and the turn to small scale investigations, the deployment of the concept has become more restricted. Statistical analysis gained a greater reputation for clarity, despite the still-not-clear appreciation of spatial relationships.

With the extension of quantitative methods in social geography, leading to quite different research aims, it later became possible not only to distinguish social groups as existent, but also to describe them in terms of combinations of characteristics. With a radical shift in approach, research became more concerned to use social statistics to distinguish groups in terms of demographic, economic, religious, and ethnic criteria, with the aim of a finer analysis of the regionalisation of social space (Steinberg, 1964).

Such aggregates of data, producing complex data patterns, were then quite freely presented as social groups, which in no way corresponded to the now old-fashioned life-form groups. As computerised pictures of social life, the technique produced numerous areal maps and descriptions of localised phenomena. What these groups based on selected characteristics lacked - and this applies also to the rather senseless further levels of aggregation - were indicators of real spatially effective social action units. Despite the increased range of characteristic groups, the new aggregating techniques for statistical materials were productive so long as they were employed with family and large units or in the analysis of process, but not when based at the level of the individual. The indicator method and process analysis were also modified by the elucidation of definite spatially effective types of household - and by the actions of these household types to activity and decision-making behaviour groups - to a useful group concept that played an essential role in discussion (Durr, 1972; Shaffer, 1968a; 1968b).

There emerged a tendency to define social geographical groups first in terms of activity space with common distance thresholds, and secondly by a group of aggregate characteristics - a tendency that could lead to ambiguity. The problem of the social geographical group nevertheless appears - rather unsatisfactorily - to have been solved up to the present (Freisl, 1976).

Social-Ecological Contribution

Measured in terms of total research product, and in terms of its methodolgical potential, the social ecology approach in German social geography has produced remarkably little. This is astonishing in view of the fact that the methods developed in the American social research school have considerably enriched the whole of German social geography. The unusually-attractive themes associated with this avenue of advance, for combined social and spatial planning, have resulted nevertheless in only a limited activity dealing with the ecologically-explicable relationship of space and society:- the questions of selective, distributive and contractive processes in space on the one hand, and the social dispersal, strain and frustration effects on the other. The questions asked in the post-war years were in a sense more mundane; one thought more immediately of the repatriation of one million refugees, or the problem of migrant workers in German growth-areas, or the commercialisation of the inner city accompanied by suburban growth. In fact only a limited number of the available themes were opened up.

Some of the grounds for this relative neglect - with of course a number of well-known exceptions (Ipsen, 1957; Mackensen & Papalekas, 1959) - were German. Other grounds for the neglect can perhaps be attributed to the tendency in the USA and in Great Britain to specialise, compartmentalise and particularise, as for example with criminal geography, or many aspects of election geography. To German eyes these were simply labelled "geography" by their originators; but were considered by German geographers to be arbitary divisions of social phenomena, part of spatial science and not geography. Further, it may well have been that social ecology found little echo in that the sort of questions it posed did not suit the political world view of the generally conservative German academic circles, and indeed appeared to offer only politically-explosive themes.

The present day enrichment of geography with themes derived from spatial analyses, still largely a separate area of study in Germany, could produce immediately considerable changes in the social ecological approach. In the meantime only a few individual studies have appeared with a social ecological approach to spatial situations - as with the social subsystems in Innsbruck (Bodzenta, 1959), competition for space in main shopping streets (Bubschmann, 1952) and commuting (Klingbeil, 1969).

The underrepresentation of this research avenue leads to the opinion that the social ecological concepts, even if not present under that name, occur implicitly as working hypotheses in other research fields.

Social Space Research

In contrast, the analysis of social structures has played a large role in German social geography. This is to be explained by the fact that ample data were accessible and suitable communities for study freely available. Furthermore the clear areal differentiation of this approach served indirectly as a large scale reference. The approach developed in three separate directions - here briefly mentioned - topographic, functional, and sociometric. In this sequence social space research mirrored the phased development of the whole discipline.

In the topographic direction various attempts were made to use a regionalisation based on statistical data to investigate the social space patterns of various areal units. In scale these attempts extended from individual households, tenement blocks and streets to towns, districts, regions and _lande_. The lack of congruence with the regions used for official statistical purposes permitted only superficial mathematical involvement in the solution of problems. A few examples can be selected: in Rhein-Main Geipel (1961) used social stratification criteria; while Braun (1968) emphasised the importance of professional differentiation in Hamburg. Ganser (1966) probed the possibilities of regionalisation based on political voting in Munich. For interregional comparisons Bobek (1968, 1970) returned to a combination of employment structure and service industries.

The social space so distinguished on the basis of selected characteristics nevertheless lacked any demonstrable social interrelationship in the form of contact, kinship or even simply recognition. There began therefore a search for criteria - from this functional viewpoint - that would help to comprehend social space as a lived-in interaction region. This was initiated by Hartke (1952) in terms of newspaper distribution. Later Scholler (1955) used membership of various societies to determine contact space; and more recently Weber (1977) has considered social space on the grounds of choice of marriage partner.

Where these functional studies could be based on official data, but as in studies of preferred space - here understood as part of the sociometric study area - new statistical methods were fully employed. And at the same time the emphasis on interviews and personal questionnaires was introduced. The spread of this approach was at first slow, but has already produced interesting results. Examples of this work which may be quoted are the studies by Monheim (1972) on the role of site preference in industrial location, Ruhl (1971) on the image of Munich, and Muske (1975) on the motives in the selection of places of higher education.

Indicator Method

In contrast to the above named perspectives of German social geography that today are either being further extended, or are in the initial testing period, the indicator method belongs generally to the past. This is true for the indicators of the cultural landscape. The analysis of non-physiognomic indices (data) remains however of permanent value (Hartke, 1959).

The attempt to understand changes in social stratification in terms of their expression in the landscape was limited - on grounds of immediacy - to the agricultural landscape. During the period of great economic restructuring in the '50s and '60s - the period when the indicator method was in use - these landscapes had been subjected to a great number of short term intensive changes. Of diagnostic value for distinguishing the contributory social factors were those changes in landuse due not to a change of cropping but to structural change in rural society. For example, reference may be made to the increase in extensive farming methods, exemplified by the increase in permanent grass, field cropping of vegetables, and fallowing. Of equal significance was reintensification, as with weekend farming and retirement farming (Ruppert, 1955, 1960). Within this range of themes is the repetitive cycle of landuse, as examined in the work for example of Hartke (1956) relating to the abandonment of tillage, i.e. "social fallow", in Frankenberger's (1959) study of afforestation, and the work of Meffert (1968) on special crops. The work of Borcherdt (1961), or at least that part of this work dealing with agricultural innovation, can be considered as within this approach.

The lack of application of the method in more recent social geography is due not only to the exhaustion of the range of phenomena to which the method was readily applicable, but also the social ambiguity of the indicators themselves. It became more and more necessary to distinguish the contribution of specific social management, leading to an unspoken tendency to re-examine the social region directly without any landscape analysis.

Process Analysis

Research into social geographical processes, their expression in changed location over space and time, is associated today with the Munich school, largely dependent upon that school for personnel and concepts, stemming from the large student-body of Hartke (Ruppert and Schaffer, 1969).

As mentioned the centre of gravity of process analysis lies in the analysis of the social <u>Grunddaseinfunktionen</u>, i.e. basic and necessary functions by which any society makes provision for its members; the spatial and temporal fulfilment of these functions having geographical expression. The spatial patterns - thus built up or destroyed - of habitation, work, provision, recreation, education and so on are shown to be in reality interwoven. But the separation of the individual processes is at first necessary in order that a selected functional field may be studied. This dissection of the corpus of functions necessary on technical grounds does not underline a priority for one or other function, but is nevertheless often so misunderstood (Leng, 1973).

Process analysis began with the study of mobility. In newly built areas it could be shown that the inner city removal corresponded with certain development stages in the household (growth, stagnation, shrinkage) (Schaffer, 1968). These stages were later employed to typify urban groups (Paesler, 1976) and in relation to the home origins of foreign workers (Schrettenbrunner, 1970). Interest has also focussed on the social-geographical aspects of leisure, particularly second homes, local recreation, and retirement homes (Ruppert & Maier, 1970; Maier, 1976). Finally the interest in the geography of education stimulated by Geipel (1965, 1970) has given a new view of the social and spatial circumstances of educational needs and of the planning of places of education, a field where both research interest and practical educational policy are closely involved.

Work on the other basic functions or processes of society has not proceeded at the same pace, and much is still awaited; but with the important conceptual changes in the methodology of social geography much may never appear.

Change in Orientation

More or less since the development of process analysis there have been increasing signs that German social geography, as here portrayed - and already to a certain extent "classical" - is clearly approaching its end, certainly in the applied sense. Some research perspectives were burnt out at the beginning of the '70s; others were not followed further, whilst some suffered a drying-up of material. The number of studies published under this heading is reducing; and a text on social geography has yet to appear. A second generation of social geographers has become unlikely - with the described concepts and methodology leading to little fundamental, new knowledge.

The turning point for the striking and rapid re-orientation was the realisation that the object "society" was too simplistlic - similar to "population" - although not so much as certain other geofactors. What has been understood under the paradigm "social area" was now seen as solely the external side of social life, the social substrate in the widest sense. Hidden in this, although only accessible through the same processes, and disclosed step by step, is the inner side of the social complex, lit by the mental complexes of social existence. This progressive expansion of research into the field of human motivation processes, preference building and decision making in their spatial aspects is labelled "environmental perception", and the title "social geography" is withheld. Although introduced and shaped by social geographers, it is in content and above all in its methodology geographical behavioural research (Thomale, 1974). Social geography has become behavioural geography.

As we ponder upon a provisional appreciation of the contribution to knowledge resulting from thirty years of social geography in Germany, it becomes clear that it has provided, as these somewhat pretentious terms would indicate, a step on the way toward a better-focussed geography of people. The interpretation of man as part of the living inventory of space - the interpretations approached by population geography - could not persist. The subsequent social geographer's view of man as an actor locked into society is thus to be seen as the necessary intermediate step, upon which could not be built an understanding of man as a spatially aware decision maker. Thus the beginnings and the later stages of the discipline in Germany form a comprehensible sequence following a certain historical research logic.

Note

This paper was originally translated by E.M. Yates of Kings College, London. For inclusion in this volume, it has been subsequently modified by the editor.

References

Bartels, D.	1968	Turkische Gastarbeiter aus der Region Izmir. Erdkunde 22, 313-324
Bobek, H.	1948	Stellung und Bedeutung der Sozialgeographie. Erdkunde 2, 118-125
Bobek, H.	1957	Gedanken uber das logische System der Geographie. Mitt. Geogr. Ges. Wien 99, 122-145
Bobek, H.	1968	Erwerbstatigenstruktur und Dienstequote, in: Ruppert, K. (Hg.) 1968a, 119-131
Bodzenta, E.	1959	Innsbruck, Eine sozialokologische Studie. Mitt. Geogr. Ges. Wien 101, 323-360
Borcherdt, C.	1961	Die Innovation als agrargeographische Regelerscheinung.

		Saarbrucken. Arb. Geogr. Inst. Saar. 6
Braun, P.	1968	Die sozialraumliche Gliederung Hamburgs. Gottingen. Weltwirtschaft. Studien 10
Claval, P.	1964	Essai sur l'Evolution de la Géographie Humaine. Cahiers de Geogr. de Besancon 12
Durr, H.	1972	Empirische Untersuchungen zum Problem der sozialgeographischen Gruppe. Kallmunz. Munch. Stud. z. Social- u Wirtschaftsgeogr. 8, 71-81
Frankenberger, R.	1959	Die Aufforstung landwirtschaftlich genutzter Grundstucke als Index fur sozialgeographische Strukturwandlungen in Oberfranken. Kallmunz. Munch. Geogr. Hefte 18
Friest, R.	1976	Sozialgeographische Gruppen and ihre Aktivitatsraume. Diss. Munchen
Ganser, K.	1966	Sozialgeographische Gliederung der Stadt Munchen auf Grund der Verhaltensweisen der Bevolkerung bei politischen Wahlen. Kallmunz. Munch. Geogr. Heft 28
Geipel, R.	1960	Erdkunde, Sozialgeographie, Sozialkinde. Frankfurt/M.
Geipel, R.	1961	Die regionale Ausbreitun der Sozialschichten im Rhein-Main-Gebiet. Bad Godersberg. Forsch. Dt. Lankeskunoe 125
Geipel, R	1965	Sozialraumliche Strukturen des Bildungswesens. Frankfurt/M, Berlin, Bonn, Munchen
Geipel, R.	1969	Industriegeographie als Einfuhrung in die

		Arbeitswelt. Braunschweig
Geipel, R.	1970	Der Beitrag der Sozialgeographie zur regionalen Bildungsforschung. Forsch. u. Sitzungsber. Akad. Raumf. u. Landsplanung Hannover 60, 1-16
Gerling, W.	1965	Kritische Bemerkungen zur Sozialgeographie. Wurzburg
Hartke, W.	1952	Die Zeitung als Funktion sozial-geographischer Verhaltnisse im Rhein-Main-Gebiet. Frankfurt/M. Rhein-Mainische Forschungen 32
Hartke, W.	1956	Die "Sozialbrache" als Phanomen der geographischen Differenzierung der Landschaft. Erdkunde 10, 257-269
Hartke, W.	1959	Gedanken uber die Bestimmung von Raumen gleichen sozialgeographischen Verhaltens. Erdkunde 13, 426-436
Hartke, W.	1963	Das Land Frankreich als sozialgeographische Einheit. Frankfurt/M., Berlin, Bonn
Hawley, A.H.	1962	Theorie und Forschung in der Sozialokologie, in: Konig, R. (Hg.) 1962-1969, 1,480-497
Hottes, K.H.	1968	Sozialgeographie. Braunschweig. Westermanns Lexikon d. Geographie 4, 339-343
Hubschmann, E.	1952	Die Zeil. Sozialgeographische Studie einer Strasse. Frankfurt/M. Frankfurter Geogr. Hefte 26
Ipsen, G.	1957	Standort und Wohnort. Okologische Studien. Koln/Opladen
Klingbeil, D.	1969	Zur sozialgeographischen Theorie und

		Erfassung des taglichen Berufspendelns. Geogr. Zeitschr. 57, 108-131
Konig. R. (Hg.)	1962-69	Handbuch der empirischen Sozialforschung. 2 Bd, Stuttgart
Leng, G.	1973	Zur Munchener Konzeption der Sozialgeographie. Geogr. Zeitschr. 61, 121-134
Mackenroth, G.	1953	Bevolkerungslehre. Berlin, Gottingen, Heidelberg
Mackensen, R. & J.U.A. Papalekas	1959	Daseinsformen der Grosstadt. Tubingen
Maier, J.	1970	Zur Geographie des Freizeitverhaltens. Kallmunz Munch. Stud. Sozial-u. Wirtschaftsgeogr. 6
Maier, J.	1976	Zur Geographie verkehrsraumlicher Aktivitaten. Kallmunz. Munch. Stud. z. Sozial-u. Wirtschaftsgeogr. 17
Meffert, E.	1968	Die Innovation ausgewahlter Sonderkulturen im Rhein-Mainischen Raum. Frankfurt/M. Rhein-Mainische Forsch. 64
Monheim, H.	1972	Zur Attraktivitat deutscher Stadte. Munchen. Ber. zur Regionalforsch. 8
Muske, G.	1975	Motive fur die Wahl des Studienortes Munchen. Kallmunz. Munch Geogr. Hefte 38
Otremba, E.	1962	Die Gestaltungskraft der Gruppe und der Personlichkeit in der Kulturlandschaft, Verh. Dt. Geographentag Koln, 33, 166-198
Overbeck, H.	1954	Die Entwicklung der Anthropogeographie (insbes. in Deutschland). Blatter z. Dt. Landesgesch. 91, 182-244

Paesler, R. 1976 Urbanisierung als sozialgeographischer Prozess. Kallmunz. Munch. Stud. Sozial-u. Wirtschaftsgeogr. 12

Paesler, R., K. Ruppert, & F. Schaffer 1977 Sozialgeographie. Braunschweig.

Ruhl, A. 1938 Einfuhrung in die allgemeine Wirtschaftsgeographie. Leiden

Ruhl, G. 1971 Das Image von Munchen als Faktor fur den Zuzug. Kallmunz. Munch. Geogr. Hefte 35

Ruppert, K. 1960 Die Bedeutung des Weinbaues und seiner Nachfolgekulturen fur die sozialgeographische Differenzierung der Agrarlandschaft in Bayern. Kallmunz. Munch. Geogr. Hefte 19

Ruppert, K. 1968a Zum Standort der Sozialgeographie. W.Hartke zum 60. Geburtstag. Kallmunz. Munch. Stud. Sozial-u. Wirtschaftsgeogr. 4

Ruppert, K. 1968b Die gruppentypische Reaktionsweite, in: 1968a, 171-176

Schaffer, F. 1968a Untersuchungen zur sozialgeographischen Situation und regionalen Mobilitat in Grosswohngebieten, a.B. Ulm. Kallmunz. Munch. Geogr. Hefte 32

Shaffer, F. 1968b Prozesshafte Perspektiven sozialgeographischer Stadtforschung, in: Ruppert, K. 1968a, 185-207

Schaffer, F. 1971 Prozesstypen als sozialgeographisches Gliederungsprinzip. Mitt. Geogr. Ges. Munchen 56, 33-52

Schaffer, R. 1969 Zur Konzeption der Sozialgeographie.

		Geogr. Rundshau 21, 205-214
Scholler, P.	1955	Einheit und Raumbeziehung des Siegerlandes. Munsten
Schrettenbrunner, H.	1970	Bevolkerungs-u. sozialgeogr. Untersuchung einer Fremdarbeitergemeinde Kalabriens. Munchen, WGI-Ber.z.Regionalforschung 5
Sievers, A.	1964	Ceylon. Gesellschaft und Lebensraum in den orientalischen Tropen: Wiesbaden
Steinberg, H.G.	1964	Fragen einer sozialraumlichen Gliederung auf statistischer Grundlage, Raumforschung und Raumordnung 22, 65-76
Storkebaum, W. (Hg)	1969	Sozialgeographie. Darmstadt. Wege der Forschung 59
Taylor, G. (Hg)	1953	Geography in the 20th Century. New York/ London
Thomale, E.	1972	Sozialgeographie. Eine disziplingeschichtl. Untersuchung zur Entwicklung der Anthropogeographie. Marburg. Marburger Geogr. Schriften 53
Thomale, E.	1974	Geographische Verhaltensforschuing. Marburger Geogr. Schriften 61, 9-30
Totten, D.E.	1959	Erdol in Saudi-Arabien. Heidelberg. Heidelberger Geogr. Arb. 4
Vogel, I.	1959	Steinkohlenbergmann - Braunkohlenarbeiter. Berichte Dt. Landeskunde 23, 215-224
Vooys, A.C. de.	1950	De ontwikkeling van de sociale geografie in Nederland. Groningen
Weber, B.	1977	Sozialraumliche Entwicklung des

		Siegerlandes. Bonn. Arb. z. Rhein. Landeskunde 43
Wirth, E.	1969	Zum Problem einer allgemeinen Kulturgeographie. Erde 100, 155-193
Wirth, E.	1977	Die deutsche Sozialgeographie in ihrer theoretischn Konzeption und in ihrem Verhaltnis zu Soziologie und Geographie des Menschen, Geogr. Zeitschr. 65, 161-187

Chapter Six

SOCIAL GEOGRAPHY IN POLAND

Bohdan Jalowiecki

Introduction

The paradox of contemporary geography lies in the growing contradiction between the inherited myth of the comprehensive character of the discipline and the growing specialisation of its individual branches. To date, geography has tended to view the natural and anthropogenic environment as a whole, an attitude which has brought many geographers to the conviction that their discipline is in itself multidisciplinary. This "epistemological imperialism" has varied consequences and, if institutionalised, acts as an obstacle to the development of the individual branches of geographic science.

It is natural to ask whether we may speak of one science or, rather, of many sciences springing from one stem and turned independent of, or at the most, complementary to, one another. The "unity of geography" continues to cause emotional controversy although the dispute seems quite futile because practice has solved the problem. On the one hand, there are highly specialised fields, such as climatology, biogeography, geomorphology, where nature is the subject of research and research methods are typical of natural sciences. They have come close to other natural sciences, such as geology or biology, in the course of their development. This does not mean that their research subject is totally independent of human activity, as is the case with physics, but human activity, destructive for the most part, does not affect the natural character of their substance.

The research subject of economic and social geography embraces man as such, the spatial forms he creates and his economic, social and cultural activity in space. Research methods in these branches of geography, their basic paradigms, explicative schemes or theories belong to the domain of social sciences and largely derive from economy, sociology and recently also psychology.

Thus the growing specialisation has caused physical geography and social geography to grow more and more distant. Further specialisation and subsequent divisions occur in the framework of each of them. Individual research problems have acquired the status of subdisciplines, and thus we have geography of settlement, industry, agriculture, culture, and so on. Other divisions will probably occur because there is almost no end to the various types of human activity. Yet divisions do not spring from the needs of science but from the inner developmental dynamism of scientific institutions where it is in the interests of the individual groups of scientists to launch a new subdiscipline and gain institutional independence. An analogous process occurs in other fields, e.g. sociology which has divided into several dozen increasingly distant subdisciplines.

These non-scientific considerations, which often lead to the pointless multiplication of "independent" disciplines, cause science much more harm than benefit. The flow of information is much hindered, which leads to information chaos, and the attitude of splendid isolation results in futile, pointless disputes.

Kenneth E. Boulding (1956) writes in this connection that crisis in present-day science results from the growing obstacles to effective communication among scholars. "Specialisation has crossed professional borders, communication is getting more and more difficult and the Republic of Learning has split into separate subcultures, among which there are only frail links". One wonders at times if science is not going to vanish within a community of hermits, each of whom, separated from others by a wall, will mumble something to himself in his own language understood only by himself. Spreading specialisation deafness causes those who should know what others know to fail to learn it for lack of suitably 'generalised ears'.

Thus, let us repeat, the current development of geography is burdened by the mythical unity of physical and social geography on the one hand, the growing specialisation and division into subdisciplines dealing with selected aspects of reality on the other. Personally, I am sure that there are two geographies, physical and social, complementary to each other. At the same time, I am convinced that there should be one social geography dealing with the process of the social formation and appropriation of space. The currently accepted division into economic, social and political geography on the one hand, and geography of settlement, agriculture, population, industry, and so on, on the other, is quite meaningless and illogical. It is assumed implicitly that human settlement is independent of economic, social and political phenomena, and can be investigated per se. Settlement geographers certainly take the above-mentioned processes into account, but they tend to seek immanent laws of the development of settlements, and to construct formalised models which do not only fail to explain reality but, on the contrary, cause it to appear more obscure.

Hence social geography should base its research on the processes of creation of spatial forms and methods of elucidating human spatial behaviour upon theories worked out by political economy and sociology. All human endeavours in space are condition by the law of production forces and relations.

Research Problems

Crisis phenomena, which became evident in Poland at the turn of the seventies, have brought out a problem which should find itself in the centre of interest of social geographers, namely, the processes of the social formation of space. In order to explain how space is formed, at least three questions should be answered: How is the process conditioned? Which spatial forms are primarily created? By whom are they created or, in other words, who are the actors representing definite interests?

Starting with these three questions, we may formulate an explicative scheme which may become, in a sense, a new paradigm of social geography in the eighties. This new paradigm may also permit one to combine the efforts of geographers dealing separately with the individual aspects of reality (e.g. settlement, agriculture, industry).

It appears that there are four factors conditioning the processes of the social formation of space (Jalowiecki, in press). (1)The social formation of space is limited by nature i.e. the natural environment. The limitations have a treble character. Firstly, natural space offers resistance and acts as an obstacle to the formation of definite forms. The level of difficulty in overcoming the obstacles depends on the accessible technical means. Secondly, natural space may be treated as a "property" and then deliberately left "free", uncultivated. The second limitation has become particularly important today, when human activity has markedly infringed upon natural resources and upset ecological balance. This accounts for the emergence of the concept of protected areas. Finally, the third limitation to the formation of space in many countries at the moment is determined by the threshold barriers of the endurance of the natural environment. Crossing the barriers creates the danger of ecological calamity (Kozlowski, 1983). Herein lies the basis for collaboration for physical and social geographers.

(2) Another factor conditioning the social formation of space is the development of technique and technology, in other words, the development of means of production, or, more broadly, the development of production forces. It is evident that throughout historical development, human potentialities in this field have grown quite remarkably, and the limit to this growth has by no means been achieved. Yet a number of disturbances springing from technical development have appeared both on the global scale and in the individual countries. They are becoming more and more dangerous to human existence on earth and lead to crisis phenomena.

(3) Yet another factor determining the social formation of space rests on the conditions of domination and dependence. In the economic aspect, these are production, hence property, relations; in the political aspect, there is power; finally, in the ideological aspect, there is the ideology of the class, i.e. strata or groups dominating in society.

(4) Finally, the last factor limiting formation of space consists of value systems and cultural categories adding up to the model of the world prevailing in the given society (Guriewicz, 1976). This notion certainly includes the dominating ideology, although it never dominates utterly and without reservations, as counter-ideologies, which participate in the social formation of space, appear and reappear. Space, to paraphrase Lotman and Uspienski (1975) is the lasting memory of society, and three types of filling it can be distinguished: quantitative enlargement of the range of space by filling its "frames" (fragments) with various texts; rearrangement of the structure of elements, as a result of which its character is altered; and

forgetting, i.e. the elimination of the individual frames and spatial forms.

The concept of spatial values is an important category in the social formation of space. Because space is society's lasting memory, its rearrangement and forgetting may be precluded in some cases, hence limiting the potentiality of its actual formation. This particular valorisation of certain sections of space is, for instance, the basis for the protection of monuments and time-honoured areas.

The categorisation of spatial forms is of key importance to the investigation of the process of the social formation of space. The categories are distinguished and classified on the basis of the role they play in production process, reproduction of labour power and re-creation of the social relations of production. Thus spatial forms are areas with a definite goal and function extending over the accompanying furnishings and material equipment. There are five basic types of spatial forms: of production, consumption, power, symbolism and exchange. Each of them naturally includes a different number of elements or subspaces. This classification draws, to an extent, upon the elements of Castells' (1975) municipal system, yet it has a much broader range.

In accordance with the basic division of labour, we divide the spatial forms of production into agricultural and non-agricultural ones. In the former case, we have systems of open space, fields and the corresponding types of farms and villages. A farm or a village is certainly more than a spatial form of production; it is also an area of consumption. In a village we also come across spatial forms of non-agricultural production, and symbolic spaces, including churches, chapels, cemeteries, crosses, and so on. Yet because of the role they place in the division of labour, both farms and villages are primarily spatial forms of agricultural production.

The space of non-agricultural production is likewise differentiated. Thus, alongside typical units of material production, such as a craftsman's workshop, a factory or a mine, there are also spaces of non-material production offering services, knowledge and information. They include offices, laboratories, etc. Hence, we include among the spatial forms of production all areas (with their accompanying furnishings) producing material goods, services and information (Castells, 1975).

Consumption space is an area of activity linked

with the social, communal and individual assimilation of a product and the surplus value, together with the simple and extended reproduction of labour power.

Thus consumption space linked with the simple reproduction of labour power includes homes with a minimal material equipment - the so-called technical furnishings. On the other hand, space linked with the expanded reproduction of labour power extends over biological systems, i.e. health services, open space and recreation areas, and over institutional systems, i.e. ones developing the individual's socialisation power, or education, some of the ideological apparatus of the state and socio-cultural institutions being linked with the ideological systems to a greater or lesser degree.

The space of power may be understood in two ways, either as territory over which its jurisdiction extends, or as the area of institutions, by means of which power is exerted. If we adopt the narrower understanding of the space of power, it will include the areas occupied by the buildings and installations employed by the state administration, police and army.

Symbolic spaces include various places and areas containing values much esteemed in the given community. These may be values linked with the joint past, such as all sorts of monuments, time-honoured buildings and quarters, and spaces of actual symbolic behaviour linked with cults or the manifestation of ideological attitudes.

Finally, the space of exchange consists of as many sub-elements as there are possible transfers in the framework of each of the formerly enumerated spaces (Castells, 1975). The space of exchange may be understood in two ways. In one understanding, it includes variously sized areas for the transfer or shift of goods, information and people. These areas may be described as "markets" on a different scale and of different types. On the one hand, they include local, regional, national and international markets; on the other, markets of goods, services, labour and information.

In another understanding, the space of exchange may be understood as an area intended for exchange, together with its furnishings. In this sense, it includes centres and districts of commerce, stores, transport fittings, such as railway stations, airports, ports, etc. The number, the quality and the localisation of fittings conditions the range of the given market, hence it is an important element of the social formation of space. To distinguish them from the space of exchange, we are going to call all sorts of network elements, together with means of transport and communication (roads, railway, telephones) carriers of exchange or connection.

The distinguished category of spatial forms does not just link up the general theory of production methods, developed on the ground of political economy and sociology with the categories of space peculiar to geography, but it also links up the individual sections of social and economic geography such as geography of agriculture, industry, services, settlement and transport.

Another important category embraces the actors of the social formation of space. Space is a social product and hence it is not generated by mysterious biological or subsocial forces, but by actual people operating in the framework of historically shaped social structures. Every space is a correlate of social classes, strata and groups. Therefore, behind every spatial form, there are actors who have created it in accordance with their interests and ideology, and whose needs it satisfies. Hence actors are a historically changeable category specific to the given production method.

During the period of monopolistic state capitalism, a meaningful change of actors occurs: huge economic organisations, with incomparably greater potentialities of shaping space than individual capitalists, come to the fore. An operation of zoning is carried out in urban space in the interest of these organisations: on the one hand, separate agglomerations of industry, vast spaces of production, come into being; on the other, huge spaces of consumption (simple reproduction of labour power) i.e. vast housing estates. One could recently observe a further stage of spatial zoning, as a result of which services and shops are isolated from housing complexes and grouped into huge centres of exchange.

At the same time, the supervising and regulating role of the state intensifies, as specialised bodies for physical planning are launched and operate for the sake of the reproduction of the existing social relations. This has led to the emergence of another specialised professional group in addition to architects, namely, town-planners or physical planners, originally economists, geographers, sociologists.

At the first stage of the socialist formation, political instances are the main actors of the social formation of space besides the state because major spatial decisions are made by the supreme party instance. Industrial and building companies also play an important, sometimes autonomous role when, availing themselves of the adopted strategy of strenuous industrialisation, they materialise their own spatial interests at the expense of other social groups.

The category of actors has introduced elements of group and institutional dynamics into our analysis, and permitted us to link the categories of social geography up with theories worked out by the sociology of politics and organisation.

The analysis proposed rests on the definition of space which, in our understanding is both a demesne (limited and unrenewable) and a field for the materialisation of the potentialities of people who generate it constructing culturally and socially, and hence also historically, conditioned forms. Shaped, socially generated space remains in a dialectical link with "free" space, space treated as a property. In the course of historical development, these relations become more and more conflicting as a result of the relative scarceness of space and the upset ecological balance.

The generated forms and types of space acquire definite meanings of a varying degree of codification, which function, as it were, independently in spatial representations. Thus people produce their space by shaping its forms which have functions and meanings. The space produced becomes the material frame of life and conditions human behaviour through the number, quality and accessibility of places where people may satisfy their needs. Yet space has a symbolic as well as a material dimension because it is permeated with emotions, feelings and values.

The symbolic space produced conditions human endeavours just like the material frame of life. Churches, cemeteries, monuments, or time-honoured areas linked with a community's history constitute, for instance, a spatial "taboo", the transgression of which is treated as sacrilege and is forbidden both by customs and the law. The double character of space results in three kinds of human behaviour in space.

Firstly, there are spatial practices conditioned by the methods of production and the prevailing ideology, in the framework of which space is shaped.

Secondly, space is marked or the individual forms are given specific meaning in accordance with the prevailing social relations and ideology. Space may be marked not only by the class (group) in authority but also by other classes and social groups in accordance with their convictions and varying degrees of verbalisation.

Thirdly, there is the process of the appropriation of space, in the course of which both individuals and social group adapt themselves to the existing space as well as adapting space to themselves and their needs by modifying and marking it. The last-named process, however, is spontaneous and non-formalised, and often also unconscious.

But, people appropriate space in continual interaction with spatial forms and other people living in the same environment (Chombart de Lauwe, 1976). They organise the space of their homes, they re-fashion and furnish it, they tread paths across the lawns in their housing estates and thus modify the existing system of passages. In some countries, foreign workpeople appropriate whole districts by gradually transforming their character so that the resulting shape and atmosphere allow them to feel "at home". Thus spaces may be divided into "alien" and "appropriated". A working class man cannot feel at home in an upper-middle class area, and neither is a bourgeois likely to feel at ease in a working-class district. These divisions reflect, therefore, deeper social differences and conflicts. The degree of the appropriation of space depends on the scope of unrestrained behaviour, on the freedom of action and the range of possible modifications.

In the formalised space of a factory or an office, a worker or an employee may only appropriate their direct working space, while the remaining space is inaccessible and alien to them to a greater or lesser degree. The situation is very well described by the workman quoted by Chombart de Lauwe (1976): "In a factory, a worker is bound to his machine: a foreman has 70m^2 at his disposal; a workshop manager has 100m^2; a factory owner 2-3 ha, a great businessman has a whole planet".

Similarly a resident of a contemporary huge housing complex has several dozen square metres of his flat at his disposal, while what begins beyond the threshold is "alien" space, "no-man's-land". On the other hand, in the traditional working-class area, stairs, landings, corridors, pavements, shops and bars added up to the jointly owned, perfectly assimilated space where intense social life went on, and the division into the private space of one's home and the public space of the street was rather inconspicuous.

At present, there are growing obstacles to the

appropriation of space. Chombart de Lauwe (1976) ascribes them to the following phenomena: a growing sense of menace resulting from overpopulation, environmental pollution and degradation of cities, the disappearance of vantage points enabling one to find one's bearings both in space and in social structure, which is caused by information chaos, and a formalised, abstract organisation of space. Thus individuals and social group are unable to adjust themselves to rapid spatial transformation prompted by technological development. Another factor which may be added to the above three is the incompatibility of space-associated culturally rooted archetypes and the actual methods in which space is generated.

Investigation of the processes of the appropriation of space which originally sprang from analyses serving as a basis for mental maps offers an important research field of social geography in the framework of which it comes close to the achievements of contemporary psychology.

Thus the theory of the social formation of space does not only act as a basis for the investigation of phenomena and processes in the framework of social geography, but it may be conducive to the integration of social geography as such, and cement its links with related field such as economics, sociology and psychology. Kuklinksi (1984) stressed the inspirational function of these disciplines leading to the creation of a new paradigm of social geography. He puts forward new questions concerning objective reality from a standpoint close to ours. When writing about the polarised development of Polish geography in the eighties, he sees the main task of social geography in research on "the geography of social crisis and social reconstruction in Poland in the perspective of international comparative studies. The new social and cultural geography of Poland in the perspective of changing system of values, changing quality of life and new trends in the development of local and regional communities" are included. In order to explain the mechanisms of social crisis and the social reconstruction of space, the processes of the social formation and appropriation of space should be investigated.

Research Methods

Here the word "methods" is used in a narrower understanding of the concept, i.e. as a set of research tools and techniques used for the penetration of reality. Research methods of sciences dealing with individuals and society are coming close to each other and interpenetrate. They are used with different competence by representatives of different disciplines. In this field, much remains to be done, beginning with the period of university studies.

In the course of their development, the individual disciplines, have worked out their peculiar methods which novices accept almost automatically. It is a truism to say that a historian has an almost ingrained ability of critical analysis of written sources, a skill that for instance a sociologist has not mastered to a sufficient degree. The latter is quite proficient in handling questionnaire techniques and interviews. Some sociologists are skillful at analysing so-called personal documents, written statements, diaries, etc. A geographer is less competent in this respect although he, too, makes use of these research tools. On the other hand, geographers are usually more proficient in techniques of the quantitative analysis of statistic data, and in using cartographic sources. A psychologist, in turn, knows subtle techniques of experimental research in small groups which may be, and in fact are, very useful in the analysis of methods of the appropriation of space.

Qualitative analysis is the basic methodological weakness of social geography. Among other things, it is perfectly evident in research on social areas in cities or the spatial-social space of cities by means of factor analysis. This rather sophisticated statistic technique brings results, the empirical sense of which is often vague. It is insufficient for the investigation of reality, for which purpose it has been used, and must be complemented with the qualitative analysis of data, which may become the source of further interpretation of phenomena under investigation.

When geographers, all in good faith, use questionnaire technique, a methodologically well trained sociologist may be taken aback by their primitive questionnaires, errors in posing questions, and interpretative carelessness resulting from a failure to understand the limitations of the technique.

There is only one solution: a better command of these techniques. A similar situation happen when a geographer tries to interpret documents and other written sources, much to a historian's dismay.

Qualitative analysis is absolutely indispensable in social geography, hence a better command of appropriate research methods, worked out by other disciplines, is also necessary. This is all the more important as the problems of the social formation of space implies the use of a whole set of methods that social geography views as unconventional, such as analyses of the actors of the social formation of space, hence organizations and institutions.

Developmental prospects of social geography, its innovative role in science and its power of elucidating increasingly complex reality depend, as Kuklinksi (1984) rightly observes, on the ability to formulate new questions and provide new answers. This paper on research problems in the eighties is a proposal in this field.

References

Boulding, K.E. 1956 General Systems Theory - the Skeleton of Science, Management Science 3

Castells, M. 1975 La question urbaine, Paris

Chombart de Lauwe, P.H. 1976 Appropriation de l'espace et changement social, in: Korosec-Serfaty (ed). Appropriation de l'espace: Actes de la conference de Strasbourg

Guriewicz, A. 1976 Kategorie kultury sredniowiecza (Categories of Medioval Culture), Warsaw

Jalowiecki, B. in press The Social Formation of Space in: B. Hamm & B. Jalowiecki (eds) The Social Nature of Space, The Hague

Kozlowski, S. 1983 Przyrodnicze uwarunkowania gospodarki przestrzennej Polski (Natural Conditions of Poland's Spatial Economy), Warsaw

Kuklinski, A. 1984 Polish Geography after World War II, Mainsprings of Development, in R.J. Johnston, P. Claval (eds): Geog-

		raphy Since the Second World War. Croom Helm, London
Lotman, J. & B. Uspienski	1975	O semiotycznym mecha nizmie kultury in: Semiotyka kultury, Warsaw (On the Semiotic Mechanism of Culture)

Chapter Seven

TOWARDS AN ANTIPODEAN SOCIAL GEOGRAPHY

Trevor R. Lee

Introduction

Any attempt to review social geography in Australia and New Zealand raises the question of whether the focus is on social geography _in_ Australia and New Zealand, or the social geography _of_ these antipodean outposts. The pedantry of such a distinction is, however, more real than apparent, especially if social geography _in_ this region is taken to mean the state of on-going intellectual debate that characterises the growth and development of a discipline or subject area. At the risk of sending the reader directly to the next chapter, it must be stated from the outset that there is little evidence that Australia and New Zealand are hot-beds of intellectual vigour in social geography. Although social geography in Australasia has tended to reflect broader, more global developments, very few social geographers in either country have been fundamentally innovative or central to such a global developments of the subject.

This disparaging assessment is not restricted solely to social geography, for a recent review of the broader field of human geography (Jeans, 1983) which drew upon Francis Bacon's distinction between 'experiments of fruit' and 'experiments of light', cast human geography in Australia and New Zealand firmly in the former category. "The first kind of investigation brings useful and practical understanding... But experiments of light, bringing theoretical knowledge to initiate and shape the experiments of fruit, are rather more difficult to find" (1983, 328).

Jeans' view of human geography in Australasia is that geographers have made significant contributions to the technical solution of problems. Much of this technical problem-orientation has been in response to the requirements of government which, in Australia and New Zealand, have probably not had the same internal research capacity as government agencies in other parts of the Anglo-American world. "It is as though human geographers ... have retreated to an Anglo-saxon pragmatism which vitiates intellectual advance, and have chosen to become technologists at the service of the state rather than an intellectual leaven in society" (1983, 328).

This is not to say that social geographers in Australasia have been unaware of the broader shifts in the theoretical focus of the subject. Indeed, they have generally embraced such trends (albeit sometimes with a time-lagged adoption) and have integrated such developments into the research of pragmatic opportunities afforded by specific social issues of national and local concern. An important question thus becomes whether social geography in Australasia has been transplanted uncritically from Anglo-American sources, or whether distinctive national themes have been identified and pursued.

The view of pragmatism as the dominant ideology has also been supported by W.B. Johnston's overview of New Zealand geography (1984). "New Zealand geographers have had to select and reject accordingly to a sense of national purpose that has developed strongly since ... 1945" (1984, 20). This sentiment, and the New Zealand geography's historic links with geology, and to a lesser extent history, are reflected in five broad themes identified by Johnston as characterising contemporary New Zealand geography" ... and they display persistent aspects of human occupance of a fragile, island environment in the South Pacific" (1984, 25). None of the five broad themes (environmental change, climate and landforms, rural change, development and under-development and 'special topics') is integrally related to social geography, although some aspects of social well-being (Kearsley, 1982), time-space concepts (Forer and Kivell, 1980) and accessibility (Barnett, 1978) are included as minor contributions to the last two themes identified by Johnston. It is, perhaps, symptomatic of the relative strengths of social geography and physical geography in New Zealand that a paper, which includes the eye-catching names of Marx and Eyles amongst its multiple authorship (Crozier et al. 1980), is related to the distribution of landslips!

But in addition to pragmatism, other explanations of the current level of debate in social geography are possible. Wynn (1984, 64), for example, has maintained "the deep-rooted anti-intellectualism characteristic of New Zealand (and other 'settler societies' in which the virtues of practical skill and pragmatic attitudes have been confirmed by material progress) retains its vitality". This theme has also been touched upon by Jeans (1983, 315). Wynn's view is that pragmatism is underdeveloped and geography can contribute much to the understanding of New Zealand society through "... the publication of good human geography beyond the 'closed corporate community' that reads academic journals. There can be no lasting refuge in empiricism". Moreover Wynn sees a strong need to make geographic research "... significant to society and to bridge the gap between the academy and the public" (1984, 67).

It is also possible to revert to elements of physical determinism to explain the lack of intellectual fervour in Australasian geography, especially the effect of the tyranny of distance on intellectual debate within a small and highly scattered geographic community. Most geography departments in Australia and New Zealand are small, very few have cells of academics pursuing similar conceptual or empirical issues. Methodological and conceptual advances are, arguably, better developed collectively through the cut and thrust of debate and through mutual and collective support for the understanding and challenging of ideas. Such environments are mostly associated with large institutions or situations where personal contact between scholars in other institutions is easily maintained. Such contact is especially difficult in the Australian context, where distances are vast and air travel expensive. Moreover, the Institute of Australian Geographers, which meets less than once every twelve months, has no social geography study group. However, as Jeans (1983, 315) points out while distance has inhibited development of a close knit community of geographers, it does not explain the lack of interaction between the several departments of geography in large cities such as Sydney.

The current orientation of social geography in Australasia is attributable then, to some combination of the elements of academic isolation and anti-intellectualism, together with forces that have encouraged pragmatic and empirical research.

While the predominant focus of social geography in Australia and New Zealand has been pragmatic and empirical, it would be wrong to conclude there have been no Australasian contributions to the development of social geography in the Anglo-American world. These contributions will be discussed in later sections of this chapter, but it is the way in which Australia and New Zealand social geography has reflected broader shifts in the orientation of the subject, tempered by specific social conditions of the two countries, that forms the major focus of this chapter. The rapid growth of geography in both nations occurred as late as the 1960s when the first waves of the post-1945 baby boom and the consequences of massive post-war immigration programmes were reflected in the growth of universities and the establishment of many new geography departments.

Today both countries are characterised by a strong dependence on primary sector activities for export earnings, but, paradoxically, are amongst the most highly urbanised nations in the world. This high level of urbanisation and the characteristics and problems of urban society, are reflected very much in the orientation of social geography in Australia and New Zealand. Social geographic studies with a rural focus are certainly found (see for example, Burnley 1976a, Chapters 4 and 5; Wood 1982) but are not common.

Although there have been some innovative subject orientation within social geography in the 1950s (notably by Scott 1955, 1964) much of the development of geography in Australia and New Zealand occurred in the 1960s as the subject was embracing the technical and methodological turmoil of quantification within the broader philosophy of logical positivism. In Australia and New Zealand these broad trends within geography were manifested most strongly within social geography through the study of social differentiation and its effects on urban residential patterns.

Social Space and Urban Residential Patterns

Although Australian urbanisation was well developed by the 1920s, there was no contemporaneous examination of Australian or New Zealand cities drawing on, much less matching, the influence of the Chicago school of urban sociology. Hence, Australasian applications of the classic models of urban social structure have generally been undertaken subsequent to the passing of major academic interest in these models. Burnley (1976, 139-145) has applied Burgess' and Hoyt's concepts to contemporary Sydney, while in an historic analysis, Lawson (1972) has used the same concepts to explore the structure of Brisbane in the 1890s. To complete the trilogy of the classic models, multiple nuclei concepts of urban structure have been related to Newcastle, N.S.W. (Daly, 1970) and to Wollongong (Robinson and Shore, 1973).

In Melbourne, Johnston (1969a) has analyzed zonal and sectoral patterns of Melbourne's residential structure and the distribution of high status groups. The stability of high status groups has also been studied by Johnston for Melbourne (1966) and Christchurch (1969b), where over time, they have not displayed the predicted drift to the periphery.

If Australasian social geography has been characterised by a dearth of studies utilising the conceptual frameworks developed by the classic North American models of urban structure, it has embraced with enthusiasm the social area analysis developed by Shevky, Williams and Bell, and its more inductive successor, factorial ecology.

Undoubtedly the bench-mark study of social area analysis in Australia is the work of Jones for Canberra (1965) and his more extensive study of Melbourne (1969). Using 1961 census data, Jones accepted the Shevky-Bell constructs "... not as a formal theory dictating the lines of subsequent analysis or urban residential differentiation, but rather as a set of sensitising concepts directing attention to basic forms of social differentiation in modern industrial society" (1969, 21). Jones used principal components analysis to transform 24 indicators of socio-economic status into three components (socio-economic status - accounting for 41 per cent of variance; zone in transition - 13 per cent; and urban fringe - 8 per cent). Similarly, he used 24 variables for household composition and 22 ethnic and religous characteristics to emulate the Shevky-Bell constructs. Jones found clearly that not only could the three Shevky-Bell components of social rank, urbanisation and segregation be identified, but in the case of each of the three sub-analyses, the Shevky-Bell dimensions were by far the most dominant ones.

Factorial ecologies of Australian and New Zealand cities have demonstrated clearly that these cities exhibit a considerable degree of social differentiation. Basic social inequalities are reflected clearly in residential patterns and socio-economic status is a dominant dimension of antipodean cities. The presence of very large immigrant communities in Australian cities ensures the emergence of the third of the three basic Shevky-Bell constructs from the numerous and more inductive factorial ecologies conducted subsequent to Jones' analysis. These studies have generally found that socio-economic status and familism are relatively independent of each other, although areas of low socio-economic status are commonly associated with immigrant populations. Ethnicity varies very much with the broad social structure of the city concerned. In Australia, Melbourne and Sydney have especially large and diverse ethnic populations with sizable southern European communities. In New Zealand, Auckland and Wellington have higher proportions of Maoris and Polynesians than do Christchurch and Dunedin (Timms, 1970).

Although variants of the three basic Shevky-Bell constructs have dominated the dimensions of urban social structure identified for different cities, (see Table 7.1), the move away from predetermined constructs which was facilitated by the application of principal components analysis to the study of social space, has unearthed some significant local variations. These include Badcock's identification of the inner Sydney life-style component he termed 'cosmopolitanism' (1973) and Houghton's identification of a specific public housing dimension in Perth (1979).

Table 7.1 Basic dimensions of urban social structure - selected Australia and New Zealand studies.

Author	Urban Area	Main dimensions
Jones (1969)	Melbourne 1961	1 Socio-economic status (SES) 2 Familism 3 Ethnicity
Timms (1970)	Auckland 1966 Wellington 1966 Christchurch 1966 Dunedin 1966	1 Familism - urbanisation 2 Social rank 3 Ethnicity
Badcock (1973)	Sydney 1966	1 Cosmopolitanism 2 SES 3 Familism 4 Migrant assimilation

Parkes (1973)	Newcastle 1966	1 Family status 2 SES (prestige) 3 Urbanism - transience 4 Ethnicity 5 Density
Stimson & Cleland (1975)	Adelaide 1971	1 Familism/growth 2 SES 3 Household structure/urbanisation 4 Ethnic structure
Houghton (1979)	Perth 1976	1 Familism 2 SES/Ethnicity 3 Mobility 4 Public housing
McDonald & Guilfoyle (1980)	Brisbane 1976	1 Life cycle 2 SES 3 Stability 4 Ethnicity

Sweetser explored the notion that there are common societal forces throughout Australia that generate systematic links between the urban residential areas. He found that similarities were more important than differences, with a critical difference between residential areas in different states occurring with the ethnicity dimension.

Unfortunately much of this work on social differentiation in Australian and New Zealand cities has been undertaken in a conceptual vacuum, and can be criticised for an overt emphasis on descriptive spatial analysis with virtually no discussion or understanding of the underlying processes. Timms suggested in an early paper on New Zealand cities that "as yet it may be doubted whether urban theory has caught up with the truth that factorial ecology can produce" (1970, 465). This problem has haunted much of this work in all parts of the world. However, it has been exacerbated in Australia by the relative lack of comparative studies, especially differences in scale of analysis, variables used and technical treatment in additional to differences arising from the specific character of each city such as its broad social or physical structure.

Two attempts to develop broad comparative research in Australia are noteworthy. The first, by Logan et al (1975), was a systematic analysis of the six Australian state capitals. Although common factors emerged, with some local variants in each city, Logan _et al_ found it difficult to construct a general neighbourhood typology for Australian cities.

The second attempt to develop a comparative analysis has been by Sweetser (1973) who has examined the social ecology of residential areas in Australia on a broad societal level rather than at the level of individual cities. Sweetser's work may well be the last word on this topic in Australia, not because all the questions have been answered and all the problems of social area analysis solved, but rather because the focus of urban and social geography in Australia has moved on.

While factorial ecologies have sought to unearth the different dimensions of urban social structure through multi-variate analysis, relatively little research has been undertaken on the individual components. The exception is the role of ethnicity and immigration and the univariate analysis of key socio-economic and demographic variables in the 25 or more social atlases which have been produced for Australian cities since the 1971 Census (see for example Davis and Spearitt, 1974; Stimson and Cleland, 1975; Houghton, 1979; McDonald and Guilfoyle, 1981; Lee, 1981). These _ad hoc_ publications for the 1971 and 1976 Censuses have been supplemented by a systematic comparative series of atlases for Australian cities at the time of the 1981 Census which has been produced by the Division of National Mapping and the Australian Bureau of Statistics in association with the Institute of Australian Geographers. In general, social atlases provide the greates single source of descriptive material on the social structure of cities. Although most of this material is highly descriptive, more recent volumes have incorporated numerous comments on equity and social polarisation (see for example, Adrian, 1983; Forster, 1984). However, they remain in Thrift's words "only a very partial instrument for understanding spatial distributions" (1983, 231).

Immigration, Ethnicity and Racial Minorities

In the period since 1945 immigration has had a major influence on the demographic and social structure of both Australia and New Zealand, and has provided a major component of metropolitan growth. In 1981, 21 per cent of Australian residents had been born overseas, and in many large cities over 25 per cent of the total population was born overseas. When the children of immigrants are added to these communities, the impact of immigration on the social structure of cities becomes even more significant. For example, in Melbourne, Sydney and Perth over half the population at the time of the 1981 Census were either immigrants, or the Australian-born children of immigrants.

Cultural ties with Britain, together with overtly racist immigration policies, coloured (or rather did not) the character of Australia's massive post-war immigration intake. The major immigrant groups to Australia have been the British, together with Italian, New Zealander, Greek, Yugoslav, German and Dutch who form a majority of the large polyglot population of Australia.

In New Zealand, British migration has also been extremely important. Immigration from southern Europe has been relatively less important than in Australia, but there has also been significant migration from Polynesia. The liberalisation of Australian immigration policies since the 1970s has increased somewhat the racial diversity of Australia's immigrant population including the entry of some 66,000 Vietnamese migrants in the period 1977-83, many entering under refugee settlement programmes.

In additional to large migrant populations both Australia and New Zealand have sizable indigenous racial minorities. Given the social, if not the numerical significance of both the Australia's Aboriginal population and New Zealand's Maori population, and the stark differences in well-being between the white populations of these two countries and their indigenous racial minorities, it is surprising how few studies have been undertaken in this area by geographers.

A paradoxical dimension of this dearth of geographic literature on racial minorities is the fact that the very first volume of the _Australian Geographer_ in 1928 included an article entitled "Some Aspects of the Aboriginal Problem in Australia" (Warner, _et al_, 1928), while some twenty years later an early volume of the fledgling _New Zealand Geographer_ carried a dauntingly similar title "Some Features of New Zealand's Racial Problem" (Meek, 1947).

Since the somewhat patronising start to geography's contribution to an understanding of each country's indigenous population, there have relatively few studies in the Australian geographic literature. Exceptions to this are Drakakis-Smith's study of access to housing by Aborigines in Alice Springs (1981) and major work by Fay Gale on urban Aborigines (1972, 1981), including the forces behind the urbanisation of Adelaide's Aboriginal population and their settlement in areas of low cost accommodation in both private and public housing. Other analysis of the urbanisation of Australian Aborigines include broad reviews (Smith 1980) as well as a more detailed attitudinal study of Queensland reserve Aborigines to cities (Guthrie, 1976).

New Zealand's geographic community has been only marginally more productive in this area. Descriptive analyses of Maori settlement have been undertaken (Metge, 1952; Frazer, 1958; Heenan, 1966), as well as Maori urbanisation (Rowland, 1971, 1972) and the interface between race and accessibility (Barnett, 1978).

However, it is the immigrant groups of Australia and New Zealand that have attracted the major research focus. Although there are many minor studies of individual immigrant groups, there has yet to be any major comprehensive geographic study of immigration and ethnicity. The emphasis on white settlement in Australia and New Zealand has given relatively little attention to immigrant racial minorities, although the Chinese have formed a small but significant group in Australia since the gold rush days of the 1850s (see for example, Wolforth, 1974; Teo, 1971).

The overwhelming focus of studies of migrant settlement has been the degree of residential segregation experienced by different groups and the basic features of their settlement patterns within metropolitan areas. The dominant techniques have been cartographic and the use of single measures of residential segregation, such as the index of residential dissimilarity, which reduces the complexities of spatial patterns to a summary statistics on a scale of no segregation (I=0) to complete segregation (I=100).

This index has generally been used, explicitly or implicitly, to test Park's assumption that "... social relations are ... frequently and inevitably correlated with spatial relations" and "physical distances so frequently are, or seem to be, the indexes of social distances" (1955). In the context of migrant settlement this nexus between spatial distance and social distance has generally seen the level of residential segregation as an index of the social processes of integration or assimilation.

The Australian research gives general support to the relationship between the social processes and spatial segregation. Such work includes studies in Queensland (Timms, 1969), Melbourne (Jones, 1967), and Sydney (Peach, 1974) which identified the greatest residential clustering amongst groups such as the Italians, Greeks and Yugoslavs.

Lesser degrees of clustering have generally been observed amongst groups such as the British, Dutch and Germans, who are less visible and whose economic integration (as reflected in occupational or income characteristics) more closely resembles that of the Australian-born population. Different levels of segregation and residential diversity are commonly related to other social indices which might reflect processes of assimilation such as language, citizenship and rates of group inter-marriage.

A doyen of the study of immigrant settlement in Australia and New Zealand has been Burnley who has undertaken a series of studies of individual ethnic groups in Australasia. These include Italians in Sydney (1981), Greeks in Sydney (1976b), German immigration and settlement in New Zealand (1973), as well as comparative studios of southern Europeans in Melbourne (1976c), and at least two general overviews of ethnic patterns in Australian cities (1974, 1976d).

The migration of Europeans to New Zealand has also been well documented (Thomson, 1967; Trlin, 1968, 1975) together with significant Polynesian migration to New Zealand which is an important component of that country's multi-racial character (see for example Pool 1961; Curson, 1968, 1970; Trlin, 1977; Gibson, 1983).

Other dimensions of ethnic migration and settlement have also been explored, including the relevance of the single "port-of entry" concept of both Australia (Lee, 1970) and New Zealand (Whitelaw, 1971; Rowland, 1971), in which overall changes in residential concentration and a general dispersal of immigrant settlement result in a weakening of the role of the inner city as an area of initial settlement for new arrivals.

The operation of chain migration has been an important process in the migrant and settlement of ethnic groups in Australia. Such migration chains involve flows of information between prospective migrants and their compatriots who have already made the move, as well as financial assistance for fares and even accommodation and employment in the areas of settlement (Jones, 1964; Price, 1965).

Relatively few of the studies of segregation have gone beyond a descriptive account (at varying degrees of sophistication) of the settlement patterns and some of the basic processes associated with changing residential patterns in the major metropolitan centres. There has been some analysis of immigration in a broader social context, such as Ware (1974) who has described southern European migration to Australia as a process of partially substituting a native-born proletariat of British ancestry with a foreign-born proletariat of Mediterranean stock.

However, the recent orientation of social geography to issues of social equity and the development of an embryonic political economy for the analysis of urban issues has not yet been reflected in changes in the character of research into ethnicity in Australia and New Zealand. Notable exceptions are works by Collins (1977) for Australia and Gibbons (1983) for New Zealand. Badcock has argued in his attempt to frame a political economy of urban space that state intervention in nineteenth century Australia was highly selective to the benefit of private capital and that "Australia governments have always viewed assisted immigration as a substitute for having to reproduce socially labour power. This policy has found the consistent support of the 'defence conscious', and private employers on the look-out for complaint construction, farm, and factory labour" (1984, 125).

A major Australian work on the concept of equity introduces the concept of the city as the spatial expression of the mode of production and the ideological construction of equity (Troy, 1981, 19). This work also explores the role of social cleavages in differential access to urban services and resources. However, one of the starkest forms of social cleavages in contemporary Australian society - the position and role of Aboriginal and immigrant groups - is virtually ignored in this critical collection of Australian essays on equity. Thus passing references only are made to migrants in terms of their dependence on out-patients facilities offered by metropolitan hospitals (Donald, 1981, 77) and their inferior housing circumstances (Neutze, 1981, 111). The failure to adequately evaluate the inequities arising from lack of access to urban resources and facilities remains a glaring inadequacy of social geographic contributions to the understanding of migrant settlement.

The sheer scale of post-war immigration to Australia focused much attention on the role of immigrants on population growth, urbanisation and industrialisation. As immigration intakes have declined increasing attention is being focused on the social consequences of Australia's transition to a multi-cultural society. The concept of multi-culturalism is one which is likely to be an increasingly significant social issue in the 1980s. Hitherto, geographic research has focused on a fairly narrow range of settlement issues and the provision of relevant geographic insights into multi-culturalism remain a challenge to be accepted by Australian geographers.

Behavioural Perspectives

The disenchantment of human geography with the shortcomings of the spatial analysis of logical positivism and the normative assumptions of neoclassical economic explanations of the spatial structure of the city is well-documented (see, for example, N. Smith, 1979).

One path from this disentchantment led to studies of perception and the decision-making process as a means of modifying the underlying behavioural assumptions of the spatial analysis approach. Early behavioural studies in New Zealand include Clark's (1969) analysis of personal information fields and Johnston's (1971) study of mental maps and suburban preference patterns. The perceived social status of different residential areas has formed another line of enquiry (Congalton, 1961; Stimson, 1972).

Another behavioural modification of the traditional spatial analysis approach has been developed by Walker (1977). Her study of urban residential differentiation in Newcastle, New South Wales, incorporated the concept of personal constructs in which the emphasis is on individually-defined dimensions of urban structure. Walker found that in no study area did the classic Shevky-Bell constructs comprise more than 45 per cent of the personal constructs elicited through conventional survey techniques.

A second stream of behavioural geography sought not a better understanding of the spatial analysis of logical positivism, but a more fundamental rejection of it. This path has led to the adoption of a greater anthropocentric perspective through humanist approaches (Ley, 1983 a, b; Gold and Goodey, 1983).

In Australia, Mercer and Powell (1972) were influential in exploring the potential of phenomenology, and their work forms a notable exception to the dearth of Australasian contributions to conceptual debate.

Despite a variety of studies covering topics such as geomancy (Yoon, 1982), humanistic descriptions of place (Jeans, 1979), and significant education essays (Hall, 1978; Fien, 1979), Jeans has commented that humanistic geography has been unable to make any significant penetration into Australasian human geography (1983, 317).

Significant contributions to the development of time-geographic approaches have been made in Australia by Parkes and Thrift (1980). Such approaches developed initially by Hagerstrand and other Swedish geographers, have provided insights into the way that time constrains individual activities and the utilisation of services.

One final area of behavioural research noteworthy in the Australian context is the behaviour of individuals in the face of natural hazards, and their perception of such hazards. The sheer size of Australia and the range of physical environments it spans have made it vulnerable to a number of recurring natural hazards, including bush fires, droughts, floods, and cyclones. The Institute of Australian Geographers and the Australian Academy of Sciences have undertaken an extensive study of natural hazards (Heathcote and Thom, 1979). This research, which reinforces the pragmatic orientation of Australian geography, includes specific studies of the perception and awareness of drought (McDonald, 1979), flood (Irish and Falconer, 1979) and bushfires (Wettenhall, 1979). In New Zealand, the list is extended to include avalanches as natural hazards (Prowse et al, 1981). In addition there have been studies of urban based environmental hazards, such as atmospheric pollution, which have been induced by people activities (Auluciems and Dick, 1976). Similar pragmatism underlines detailed studies of human responses to the effect of a bridge collapse which severed the only cross-river link between two parts of a metropolitan centre (Lee and Woods, 1981).

Poverty and Social Well-being

The myth of Australia as the 'lucky country' - an egalitarian society free of poverty and other social ills - was developed and perpetuated easily during the long post-war boom in the 25-30 years following the end of world war two. The death knell of this myth was finally sounded in 1975 with publication of the Australian Government Commission of Inquiry into Poverty (Henderson Report) - its content having been foreshadowed by the findings of an earlier survey of poverty in Melbourne in the 1960s (Harper, 1967; Henderson et al, 1970).

The Henderson Report established that 10.2 per cent of Australian households were living below the poverty line, with a further 7.7 per cent of households in a broader band of poverty within 20 per cent of the poverty lines.

The Henderson Report itself contained only limited information on the geographic distribution of poverty, although it did draw attention to broad regional differences, with the states of Queensland and Tasmania having the highest proportions below the poverty line. To a large degree, this reflects the greater proportion of the population of these states in rural areas and small towns where economic opportunities are limited, especially for women, and the economic base is commonly very narrow.

Manning (1976) has used the data from the Australian Government Commission of Inquiry into Poverty to analyse in more detail the geographic distribution of poverty between and within Australian capital cities. The affluent national capital of Canberra was shown to have the lowest level of poverty, although Adrian (1984) has identified significant pockets of poverty and deprivation that have emerged in the decade following the publication of the Inquiry's report. Poverty in New Zealand has not been subject to the same detailed analysis that the Henderson Report provided for Australia, although general analyses of income distribution and poverty have been undertaken (MacRae, 1977).

The rediscovery of poverty in the 1970s marks a significant watershed in the focus of orientation of social geographic research in Australia. Very little research undertaken since the mid 1970s falls into the mould of pure spatial analysis. Rather, the concern for social relevance, which permeated Anglo-American geography from the late 1960s and early 1970s, became increasingly reflected in the work of Australian and New Zealand social geography. Nevertheless, Australian geographers have concentrated very much on, in Jeans' terms, 'experiments of fruit' through provision of technical solutions to problems. The contemporary 'experiments of light' - specifically the rise of marxist analysis and the development of a political economy - that were spawned elsewhere in the world by the concern for social relevance received only limited attention in Australasia in the 1970s. The increasing adoption of such perspective will be discussed later in this chapter. The blend of pragmatism and social relevance that arose in the 1970s is reflected most clearly within Australian social geography in the development of studies of social well-being, accessibility, and the spatial inequalities in the provision of, and access to scarce urban resources, services and facilities.

Social Indicators and Social Statistics

Social planning in Australia has generally suffered from an inadequate information and data base from which to make judgements of social need. A number of studies have utilised single variables (usually census data) as broad surrogate measures of social problems (see for example, New South Wales Planning and Environment Commission, 1978). In addition, many of the numerous social atlases have linked the distribution of critical variables to broad social planning considerations.

Multi-variate analyses of community well-being that have been developed extensively by geographers in other parts of the Anglo-American world (Smith 1973, Coates et al, 1977) have also been adopted in Australasia, including a regional study in NSW (Vinson and Homel, 1976) and, in a different context, for development planning in New Zealand (Cant et al, 1983). Such analyses can be directed either to the identification of broad dimensions of well-being and life quality, or they can utilise a narrower set of variables to evaluate the social needs of different communities for specific facilities such as child-care (Social Welfare Commission, 1975). With minor exceptions (Kearsley, 1982) perceptual, or subjective, social indicators have been relatively underdeveloped in Australasia.

Most studies of well-being have depended heavily upon census data, although the social statistics of the judicial system, health and child welfare authorities produce very different dimensions of well-being from those derived from census material. This has been clearly demonstrated in a study of social dysfunction in Melbourne (Melbourne & Metropolitan Board of Works, 1974). The dearth of data from welfare organisations and government departments in research of this nature partly reflects the difficulty social researchers in Australasia have had in gaining access to the data sources of bureaucracies, despite the alleged pragmatic orientation of much Australasian social research.

The publication of critical social welfare data has begun to emerge, however notably with the publication of community profiles produced for Tasmania by the Commonwealth Department of Social Security. These profiles, modelled somewhat on the lines of social atlases, combine traditional census data with the government's own welfare data, such as the number and types of different pensions and benefits paid in different communities, and the duration of payment of unemployment benefits (DSS, 1984).

The Australasian geographic contribution to the social indicators literature confirms very much to the pragmatism apparent for much of antipoden social geography. There has been a technical appraisal of the choice of spatial frame for regional patterns of well-being in New Zealand (McCracken, 1983), but dearth of conceptual debate within Australasian geography related to the social theory upon which social indicator studies might be based. A notable exception to this is the work of Ife (1980, 1983) who has developed a model of social need statements. Ife has argued that territorial social indicators, and other statements of social need, are not objective and value-free statements as the positivist tradition would have it. Rather, needs statements are highly normative and value laden and, as such, it is more appropriate to look at a model of need statements in which the values and interests of those defining the needs are explicitly recognised (Ife, 1983).

From Social Welfare to Political Economy

Studies of accessibility and the distribution of urban resources epitomise the tendency for Australasian social geography to reflect the convergence of global trends in the subject, with distinctive national features and social conditions. The Anglo-American geographic community was introduced to these concept of welfare and resource allocation mainly through the works of Pahl (1970) and Harvey (1973). Such introductions included the notions of urban managerialism and other Weberian sociological concepts, as well as a focus on externalities and group conflict to manipulate 'goods' and 'bads' in cities.

The Australian urban environment has provided a fertile ground for the germination of studies on access to facilities and the equity of resource distribution. Rapid urban growth in the 1950s and 1960s created large sprawling low density suburbs in which the burgeoning population growth was not matched by government investment in basic infrastructure, nor by the adequate development of health, educational, and recreational facilities or job opportunities. In the Australian national elections of 1972 planning and public investment in cities became critical issues and formed an important part in the platform of the Australian Labour Party. Thus, it was Australian politicians, rather than academics, who were primarily responsible for raising awareness of the resource allocation. In his policy speech for the 1972 election, the successful Labour leader, Mr Gough Whitlam, made an unequivocal statement that was the precursor of much social geographic research in the 1970s:

> ... no amount of wealth redistribution through higher wages or lower taxes can really offset the inequalities imposed by the physical nature of the cities. Increasingly, a citizen's real standard of living, the health of himself and his family, his children's opportunities for education and self-improvement, his access to employment opportunities, his ability to enjoy the nation's resources for recreation and culture, his ability to participate in the decisions and actions of the community are determined not by his income, not by the hours he works, but by where he lives.

The precedence that this statement gives to the locational basis of inequality, rather than to more fundamental structural causes, was generally reflected in social geographic research during the 1970s. It was almost a decade after Whitlam's statement before the relative significance of location and structural bases of inequality was considered widely.

Thus, it was in this context of broader developments in social geography in Britain and North America in the 1970s, and increased political attention on urban inequalities that Australian social geography embraced questions of accessibility. In addition to general studies of accessibility (Black, 1977; Morris, 1978; Morris _et al_, 1978; King,

1978) there were specific studies on a variety of urban resources, including health facilities (Morris, 1976; Barnett and Newton, 1977) and other aspects of medical care (Freestone, 1975; Morris and Donald, 1980; Stimson, 1981); child-care facilities (Freestone, 1977); employment opportunities (Black, 1978; Forster, 1978); educational facilities (Walker, 1979); fire service provision (Adrian, 1980) and legal services (Vinson et al, 1977).

Collectively, these studies provide impressive documentation of the inequalities inherent in Australian urban centres. The social welfare approach to the analysis of space, reflected most clearly in Australia through studies of accessibility, marks the philosophical and conceptual transition to a new orthodoxy. Thus, the spatial analysis of the logical positivists, with its overt concern for pattern which dawned in the mid 1960s was, a decade later, entering its twilight years. In 1978 Logan foreshadowed.

> ... that human geography will join the new convergence in the social sciences and its primary aim will be ... the determination of spatial variations in human welfare. Such a human geography would be explicitly normative and based on a conflict view of society; its purpose would be interventionalist and reformist; its content focus the spatial inequalities of an increasingly urbanised society (1978, 11).

It is clear that social geography in Australia has become part of an important convergence in the social sciences in which Stretton's early work (1970) was clearly a vanguard. Much of the interdisciplinary work in social issues in Australian cities has been conducted by workers in the Urban Research Unit of the Australian National University. Much of this work was focused on questions of equity and social justice (Troy, 1981) as well as specific studies on housing (Williams, 1983a) and the political nature of the bureaucracy (Lloyd and Troy, 1981).

Thus, social geography has moved, or has perhaps been swept along with broader developments in the social sciences towards an approach of political economy. This trend has developed in Australia to a far greater extent than it has in New Zealand. While a number of marxist concepts are incorporated in this approach, much of the emphasis has been theoretical. Rather, it has focused on a broader concept of political economy of space. Thus, in their attempt to develop an Australian urban political economy, Sandercock and Berry (1983, xi) adopt as a general definition the question "who's getting what out of the urban economy and the urban planning system, and why?"

A critical element of this approach in Australia has been evaluation of how much emphasis shall be placed on spatial considerations once the structural basis of inequality is recognised. This issue, which Badcock (1984) terms 'putting space in its place', has not been widely formulated in Australia along strict Marxist lines, which would repudiate the influence of space. The spatial basis of inequality in Australian cities is increasingly recognised as reflecting more fundamental, underlying causes of inequality, but space is still credited with a significant role. Thus, "... physical space is an essential medium through which more profound causes operate (and) ... physical space compounds problems originating in more fundamental causes" (Jones and Stilwell, 1983, 25-26). Similarly, Badcock's significant formulation of a political economy of urban space looks closely at Marxist analysis but does not identify itself as such. Thus "... the most damaging and persistent inequalities are structural, whereas space functions, at best, to reinforce or mediate structually determined inequalities" (Badcock, 1984, 53). Similarly, a major Australian work on urban equity argues that while the city may be the spatial expression of the mode of production and consumption, such a simple analysis fails to address many questions, such as the extent to which state intervention can effect a more equitable distribution of resources and amenities (Troy, 1981, 19).

Urban social geography in Australia appears ripe to develop and extend analysis from the perspective of political economy. For example, the key area of housing has hitherto been examined primarily from a descriptive perspective (Neutze, 1977) or with a specific policy orientation (eg Hancock and Burke, 1983). The Australian and New Zealand housing situations, have much in common, especially the predominance of owner occupation in pattern of tenure and in household aspiration.

Weberian managerialist perspectives of housing having been explored only scantily (see for example,

Lee, 1977), while more recently there have been some attempts to set housing issues in a broader theoretical context incorporating marxist perspectives (eg Sandercock, 1979; Berry, 1983a, b; Kemeny, 1983; Milligan, 1983). There has also been some embryonic analysis of other dimensions, such as gender, which hitherto have been largely ignored (Allport, 1983).

Conclusion

Social geography does not have a long history in Australia and New Zealand, and almost all work has taken place in the last twenty years. Relatively few Australian contributions have been critical to the conceptual and theoretical developments of the subject on an international scale, yet Australian and New Zealander social geographers have adapted readily to developments elsewhere - especially from Britain and North America. Thus, the changing emphasis of Australasian social geography from the spatial analysis of logical positivism in the 1960s, to the social welfare concerns of the 1970s, and an increasing emphasis on political economy in the 1980s, is a common evolutionary path. Although it has not travelled alone on this path, Australasia has generally been a follower of change rather than a leader.

In tracing the progress of Australasian social geogaphy along the path from logical positivism to political economy, it is possible to identify an emerging local emphasis. The replication of factorial ecologies and other overt concerns with spatial pattern in the 1960s and early 1970s represents the most uncritical local adoption of social geographic approaches developed elsewhere. Studies of immigration and the ethnic dimension of urban residential structure represent local responses to pertinent social and demographic developments in both nations.

Similarly, the general social welfare emphasis of Australasian social geography in the 1970s reflected the 'relevance' and welfare concerns of social geography in Britain and North America. However, the rapid urban growth of low density, underserviced suburban areas, especially in the largest Australian metropolitan centres, gave studies of accessibility and the distribution and allocation of urban services and facilities both intellectual appeal and pragmatic value.

Australasia's first steps towards perspectives of political economy represent for social geography, a conceptual approach which is the antipathy of the unfettered spatial analysis of logical positivism. This trend is closely associated with increasing cross-disciplinary interest from sociologists, economics, historians and political scientists in social and urban questions of interest also to geographers and urban planners.

Parallelling this move to the conceptual antipodes has been an increasing awareness of the need to develop appropriate local models of concepts and approaches developed elsewhere. Examples of this growing awareness are evidence in Kendig's analysis of Australian inner city areas (1979) and in the first volumes of the Urban Studies Yearbook emanating from the ANU's Urban Research Unit (Williams, 1983b, 1984). Thus, Williams comments that "... many place specific questions are transposed between countries wiithout modification or even comment ... The pressure to conform and to replicate rather than to contrast and challenge leads both to bad theory and a weak grasp of process" (1983, 3-4). Similarly, Badcock (1984, 168) has shown how

> ... similar capitalist institutions and mechanisms do not always make for common urban outcomes; ... that government policy may give rise to indigenous expressions of essentially the same process; ... that different local conditions commend alternative patterns of investment.

If the growing awareness of these local issues continues to converge with the growing acceptance of political economy in Australasia, the prospects for development of a truly antipodean social geography seem bright.

References

Adrian, C.J.	1980	Fire service provision: a conceptualisation of demand and supply, Proceedings, IAG Conference, Newcastle, N.S.W.
Adrian, C.J.	1983	Canberra ... a social atlas, Atlas of population and housing, 1981 Census, Vol. 1, Division of National Mapping and Australian Bureau of Statistics in

		association with Institute of Australian Geographers, Canberra
Allport, C.	1983	Women and suburban housing: post-war planning in Sydney 1943-1961, in Williams, P. (ed) 1983b *op. cit*.
Auluciems, A. & J.H.A. Dick	1976	Factors in environmental action: air pollution complaints in Brisbane, *Australian Geographical Studies* 14, 59-67
Australian Government Commission of Inquiry into Poverty	1975	*Poverty in Australia: First main report* AGPS, Canberra
Badcock, B.A.	1973	The residential structure of metropolitan Sydney, *Australian Geographical Studies* 11, 1-17
Badcock, B.A.	1984	*Unfairly structured cities*, Blackwell, Oxford
Barnett, J.R.	1978	Race and physician location: trends in two New Zealand urban areas, *New Zealand Geographer*, 34, 2-12
Barnett, J.R. & P. Newton	1977	Intra-urban disparities in the provision of primary health care, *ANZ Journal of Sociology*, 13, 60-68
Berry, M.	1983a	Posing the housing questions in Australia: elements of a theoretical framework for a Marxist analysis of housing, in Sandercock, L. & Berry, M. op. cit.
Berry, M.	1983b	Whose city? The forgotten tenant, in Sandercock, L. & Berry, M. op. cit.
Black, J.	1977	*Public inconvenience: access and travel in seven Sydney suburbs*, Urban Research Unit, The Australian National

		University, Canberra
Black, J.	1978	Changes to employment accessibility in Sydney, Papers, 3rd Meeting, ANZ Section, Regional Science Association, Melbourne
Burnley, I.H.	1973	German immigration and settlement in New Zealand, New Zealand Geographer, 29, 45-63
Burnley, I.H.	1974	The social ecology of immigrant settlement in Australian cities, in Burnley, I.H. (ed) Urbanisation in Australia: the post-war experience, Cambridge University Press, Cambridge
Burnley, I.H.	1976	Social environment: a population and social geography of Australia, McGraw Hill, Sydney
Burnley, I.H.	1976b	Greek settlement in Sydney, 1947-1971 Australian Geographer, 14, 200-213
Burnley, I.H.	1976c	Southern European populations in the residential structure of Melbourne, Australian Geographical Studies, 14, 116-132
Burnley, I.H.	1981	Italian settlement in Sydney, 1920-78, Australian Geographical Studies, 19, 177-194
Cant, R.G. et al (eds)	1979	Social indicators for development planning in New Zealand, Commission for UNESCO and New Zealand Development Council, Wellington, NZ
Clark, W.A.V.	1969	Information flows and intra-urban migration, Proceedings, Association of American Geographers, 1, 38-42
Coates, B.E., R.J. Johnston	1977	Geography and inequality Oxford Univer-

& P.L. Knox		sity Press, London
Collins, J.	1975	The political economy of post-war immigration, in Wheelwright, E.L. and Buckley, K. (eds) Essays in the political economy of Australian capitalism, vol. 1. ANZ Book Co, Sydney
Congalton, A.	1961	Status ranking of Sydney suburbs, Studies in Sociology, No. 1 University of N.S.W., Kensington N.S.W.
Crozier, M.J., R.J. Eyles, S.L. Marx, J.A. McConchie & R.C. Owen	1980	Distribution of landslips in Wairarapa Hill Country, New Zealand Journal of Geology and Geophysics, 23, 575-586
Curson, P.H.	1968	Some demographic aspects of Cook Islanders in Auckland, Proceedings, Fifth NZ Geography Conference, 67-75
Curson, P.H.	1970	Polynesians and residence in Auckland, New Zealand Geographer, 26, 162-173
Daly, M.T.	1970	The development of the urban pattern of Newcastle, Australian Economic History Review, X, 190-203
Davis, J.R. & P. Spearitt	1974	Sydney at the Census 1971: a social atlas. Summitt Press, Urban Research Unit, ANU, Canberra
Department of Social Security (DSS)	1984	Regional office community profile, Tasmania 1984. AGPS, Canberra
Donald, O.D.	1981	Medical services, in Troy, P.N. (ed) op. cit.
Drakakis-Smith, D.	1981	Aboriginal access to housing in Alice Springs, Australian

Geographer 15, 39-51

Fien, J. 1979 Towards a humanistic perspective in geographical education, Geographical Education, 3, 407-421

Forer, P. & H. Kivell 1980 Space time budgets, public transport and spatial choice, Environmental and Planning A 13, 497-509

Forster, C. 1978 Accessibility of employment and the journey to work in metropolitan Adelaide, Papers, 3rd Meeting, ANZ Section Regional Science Association, Melbourne

Forster, C. 1984 Adelaide ... a social atlas, Atlas of population and housing 1981 Census, Vol. 5, Division of National Mapping and Institute of Australian Geographers, Canberra

Frazer, R.M. 1958 Maori land and Maori population in the far north, New Zealand Geographer 14, 19-31

Freestone, R. 1975 On urban resource allocation: the distribution of medical practitioners in Sydney. Geography Bulletin, 7, 14-25

Freestone, R. 1977 Provision of child care facilities in Sydney, Australian Geographer, 13, 318-325

Gale, F. 1972 Urban Aborigines ANU Press, Canberra

Gale, F. 1981 Aborigines: adjustments of migrants in cities; with particular reference to Adelaide, in Jones, G. & Richter, H. (ed) Population mobility and development, Southeast

		Asia and the Pacific, Development Studies Centre Monograph 27, ANU, Canberra
Gibson, K.D.	1983	Political economy and international labor migration: The case of Polynesians in New Zealand, New Zealand Geographer 39, 29-42
Gold, J.R. & B. Goodey	1983	Behaviourial and perceptual geography. Progress in Human Geography, 7, 578-586
Guthrie, G.	1976	Attitudes of Queensland reserve Aborigines to cities, Australian Geographical Studies 14, 43-48
Hall, R.	1978	Teaching humanistic geography, Australian Geographer 14, 7-13
Hancock, L. & T. Burke	1983	Youth housing policy, AHRC Project No. 123 Australian Housing Research Council, Canberra
Harper, R.J.A.	1967	Survey of living conditions in Melbourne - 1966 Economic Record, 42, No. 102, 262-288
Harvey, D.	1973	Social justice and the city Edward Arnold, London
Heathcote, B.L. & B.G. Thorn (eds)	1979	Natural hazards in Australia, Australia Academy of Sciences, Canberra
Heenan, L.D.B.	1966	The changing South Island Maori population, New Zealand Geographer 22, 125-165
Henderson, R.F. et al	1970	People in poverty: a Melbourne survey. Cheshire, Melbourne
Houghton, D.S.	1979	Perth at the 1976 Census: a social atlas, Department of Geography, University of Western Australia, Perth

Ife, J.W.	1980	The determination of social need - a model of needs statements in social administration, Australian Journal of Social Issues, 15, 92-107
Ife, J.W.	1983	The determination of social need. Unpublished Ph.D. thesis, University of Tasmania
Irish, J.L. & B. Falconer	1979	Reaction to flood warning, in Heathcote, R.L. and Thorn, B.G. (eds) op. cit.
Jeans, D.N.	1979	Some literary examples of humanistic descriptions of place, Australian Geographer 14, 107-214
Jeans, D.N.	1983	Experiments of fruit and experiments of light: human geography in Australia and New Zealand, Progress in Human Geography 7, 313-343
Johnston, R.J.	1966	The location of high status residential areas, Geografiska Annaler, 48B, 23-35
Johnston, R.J.	1969a	Zonal and sectoral patterns in Melbourne's residential structure: 1961, Land Economics, 45, 463-467
Johnston, R.J.	1969b	Processes of change in the high status residential districts of Christchurch, 1951-1964, New Zealand Geographer, 25, 1-25
Johnston, R.J.	1971	Mental maps of the city: surburban preference patterns, Environment and Planning 3, 63-72
Johnston, W.B.	1984	An overview of geography in New Zealand to 1984, New Zealand

		Geographer, 40, 20-33
Jones, E. & F. Stilwell	1983	When is an urban problem not an urban problem? In Williams, P. (ed) op. cit.
Jones, F.L.	1964	Italians in the Carlton area: the growth of an ethnic concentration, The Australian Journal of Politics and History, 10, 83-95
Jones, F.L.	1965	A social profile of Canberra, 1961, ANZ Journal of Sociology, 1, 107-120
Jones, F.L.	1967	Ethnic concentration and assimilation: an Australian case study, Social Forces, 43, 412-423
Jones, F.L.	1969	Dimensions of urban social structure: the social areas of Melbourne, Australia. ANU Press, Canberra
Kearsley, G.W.	1982	Subjective social indicators and the quality of life in Dunedin, New Zealand Geographer, 38, 19-24
Kemeny, J.	1983	The great Australian nightmare: a critique of the home ownership idology Georgian House, Melbourne
Kendig, H.	1979	New life for old suburbs George Allen and Unwin, Sydney
King, R.J.	1978	Accessibility and behaviour in twenty areas of Melbourne, Internal Report AIR 268-3 Australian Board Research Board, Melbourne
Lawson, R.	1972	Brisbane's population in the 1890s: an ecological study, ANZ Journal of Sociology, 8, 32-45
Lee, T.R.	1970	The role of the ethnic

community as a reception area for Italian immigrants in Melbourne, Australia, International Migration, 8, 50-64

Lee, T.R. 1977 Choice and constraints in the housing market: the case of one-parent families in Tasmania, ANZ Journal of Sociology, 13, 41-46

Lee, T.R. 1981 A social atlas of Hobart Department of Geography, University of Tasmania

Lee, T.R. & L.J. Wood 1981 Adjustment in the urban system: the Tasman Bridge collapse and its effects on metropolitan Hobart, Progress in Planning 15, 63-150

Ley, D. 1983a A social geography of the city Harper and Row, New York

Ley, D. 1983b Cultural/humanistic geography, Progress in Human Geography 7, 267-275

Lloyd, C.J. & P.M. Troy 1981 Innovation and reaction: the life and death of the Federal Department of Urban and Regional Development, George Allen and Unwin, Sydney

Logan, M.I. et al 1975 Urban and regional Australia: Analysis and policy issues Sorrett, Melbourne

Logan, W.S. 1978 Post-convergence political geography - death or transfiguration, Monash Publications in Geography 18

MacRae, J. 1977 Income distribution and poverty in new Zealand, in Pitt, D.C. (ed) Social class in New Zealand. Longman, Auckland

Manning, I.	1976	The geographic distribution of poverty in Australia, Australian Geographical Studies 14, 133-147
McCracken, K.W.J.	1983	Dimensions of social well-being: implicatons of alternative spatial frames, Environmental and Planning A, 15, 579-592
McDonald, G.T. & M.J. Guilfoyle	1981	Urban social Atlas of Brisbane - 1976 Institute of Applied Social Research, Griffith University, Brisbane
McDonald, N.S.	1979	Hazard perception in northern New South Wales, in Heathcote, R.L. & Thom, B.G. editors op. cit.
Meek, R.L.	1947	Some features of New Zealand's racial problem, New Zealand Geographer, 3, 75-82
Melbourne & Metropolitan Board of Works	1974	Social dysfunction and relative poverty in Melbourne, Advocate Press, Melbourne
Mercer, D.C. & J.M. Powell	1972	Phenomenology and related non-positivistic viewpoints in the social sciences, Monash Publications in Geography, 1
Metge, J.	1952	The Maori population of northern New Zealand, New Zealand Geographer 8, 104-124
Milligan, V.	1983	The state and housing: questions of social policy and social change, in Graycar, A. (ed) Retreat from the welfare state George Allen and Unwin, Sydney
Morris, J.M.	1976	Access to community health facilities in Melbourne Papers of the First Australia-New

		Zealand Regional Science Conference, Brisbane, 143-173
Morris, J.M., P.L. Dumble & M.R. Wigan	1978	Accessibility indicators for transport planning, Australia Road Research Board Report No. AIR 1058-3 Australia Road Research Board, Melbourne
Morris, J.M.	1978	Investigation of accessibility and mobility in urban communities, Australia Road Research Board Report No. AIR 1058-3 Australia Road Research Board, Melbourne
Neutze, M.	1977	Urban development in Australia George Allen and Unwin, Sydney
Neutze, M.	1981	Housing, in Troy, P. Ed. op. cit.
New South Wales Planning & Environment Commission	1978	Social indicators, Research study No 4 N.S.W. Planning and Environment Commission, Sydney
Pahl, R.	1970	Whose city? Longman, London
Park, R.E.	1955	The collected papers of Robert Ezra Park, Vol III ed. E.C. Hughes et al. The Free Press, Glencoe, Ill
Parkes, D.N.	1973	Formal factors in the social geography of an Australian industrial city, Australian Geographical Studies, 11, 171-200
Parkes, D.N. & N. Thrift	1980	Times, spaces and places: a chronogeographic perspective Wiley, New York
Peach, G.C.K.	1974	Ethnic segregation in Sydney and intermarriage patterns, Australian Geographical Studies, 12, 219-229
Pool, D.I.	1961	Maoris in Auckland,

Journal of the Polynesian Society, 70, 43-66

Pride, C.A. 1965 Southern Europeans in Australia Oxford University press, Oxford

Prowse, T.D. et al 1981 Avalanche as a natural hazard in New Zealand, New Zealand Geographer, 37, 25-31

Robinson, R. & J. Shaw 1973 Density gradients and urban shape in the Illawarra Corridor, Australian Geographical Studies, 11, 211-117

Rowland, D.T. 1971 Maori migration to Auckland, New Zealand Geographer, 27, 21-37

Rowland, D.T. 1972 Processes of Maori urbanisation, New Zealand Geographer, 28, 1-22

Sandercock, L. & Berry, M. 1983 Urban political economy: the Australian case. George Allen & Unwin, Sydney

Scott, P. 1964 The cinema in social geography, New Zealand Geographer, 20, 138-152

Smith, D.M. 1973 The geography of social well-being in the United States McGraw-Hill, New York

Smith, L.R. 1980 New black town, or black new town: the urbanization of Aboriqines, in Burnley, I.H., Pryor, R.J. & Rowland, D.T. (editors) Mobility and community changes in Australia University Queensland Press, St. Lucia

Smith, N. 1979 Geography, science and post-positivist models of explanation, Progress in Human Geography, 3, 356-383

Social Welfare Commission 1975 Family and child care project: needs data by

		<u>regions</u>, Statistical Bulletin, No. 1. AGPS, Canberra
Stretton, H.	1970	<u>Ideas for Australian cities</u>, Georgian House, Melbourne
Stimson, R.J.	1972	Perceived status and residential desirability of Adelaide suburbs, Paper presented to section 21, 44th ANZAAS Congress, Sydney
Stimson, R.J.	1981	The provision and use of general practitioner services in Adelaide, Australia: application of tools of locational analysis and theories of provider and use spatial behaviour, <u>Social Science and Medicine</u>, 15D 27-44
Stimson, R.J. & E.A. Cleland	1975	<u>A socio-economic atlas of Adelaide: an analysis of the 1971 Census</u>, School of Social Sciences, Flinders University and the S.A. Department of Community Welfare, Adelaide
Sweetser, F.L.	1982	<u>Urban residential areas in Australia</u>, Monograph No. 4, Department of Sociology, RSS, ANU
Thomson, K.W.	1967	Dutch migrants in the economy of New Zealand, <u>New Zealand Geographer</u>, 23, 95-105
Thrift, N.	1983	Social mapping: a partial instrument, in Williams, P. ed, <u>op. cit</u>.
Timms, D.W.G.	1969	The dissimilarity between overseas-born and Australian-born in Queensland: dimensions of assimilation, <u>Sociology and Social Research</u>, 53, 363-374

Timms, D.W.G. 1970 Comparative factorial ecology; some New Zealand examples, _Environment and Planning_, 2, 455-467

Trlin, A.D. 1968 Yugoslav settlement in New Zealand, 1890-1961, _New Zealand Geographer_, 24, 1-22

Trlin, A.D. 1975 Dutch immigrants in Auckland: a factorial ecology, _New Zealand Geographer_, 31, 124-141

Trlin, A.D. 1977 Samoan immigrants in Auckland: a factorial ecology, _ANZ Journal of Sociology_, 13, 152-160

Troy, P. (ed) 1981 _Equity in the city_, George Allen and Unwin, Sydney

Vinson, T. & R. Homel 1976 _Indicators of community well-being_, AGPS, Canberra

Vinson, T. et al 1977 Territorial justice in access to legal services, in Tucker, K.A. (ed) _The economics of the Australian service sector_ Croom Helm, London

Walker, S.R. 1977 Personal constructs and urban residential differentiation, _ANZ Journal of Sociology_, 13, 212-218

Walker, S.R. 1979 Inequalities in education and educational services, _Australian Geographical Studies_, 17, 175-192

Ware, H. 1974 The social and demographic impact of international immigrants on Melbourne: a study of various differentials, in Burnley, I.H. (ed) _Urbanisation in Australia - the post-war experience_, Cambridge University Press, Cambridge

Warner, W.L. et al 1928 Some aspects of the Aboriginal problem in Australia, Australian Geographer, 1, 67-68

Wettenhall, R.L. 1975 Bushfire disaster: an Australian community in crisis, Angus and Robertson, Sydney

Whitelaw, J.S. 1971 Migration patterns and residential selection in Auckland, New Zealand, Australian Geographical Studies, 9, 61-76

Williams, P. 1983a The politics of property: home ownership in Australia: Paper presented to Urban Research Unit seminar, Australian National University, Canberra

Williams, P. (ed) 1983b Social process and the city, Urban Studies. Yearbook No. 1, George Allen and Unwin, Sydney

Williams, P. (ed) 1984 Conflict and development, Urban Studies Yearbook No. 2 George Allen and Unwin, Sydney

Wolforth, J. 1974 Residential concentration of non-British minorities in 19th century Sydney, Australian Geographical Studies 12, 207-18

Wood, L.J. 1982 Cost savings in school transport: A Tasmanian study. Australian Geographical Studies, 20, 219-230

Wynn, G. 1984 Human geography in a changing world, New Zealand Geographer, 40, 64-69

Yoon, H.K. 1980 The image of nature in geomancy, Geojournal, 4, 341-348

Chapter Eight

SOCIAL GEOGRAPHY UNDER APARTHEID

C.M. Rogerson and J.G. Browett

The landscape for social geography research in South Africa is one which is particularly 'unsocial' in character. Since its election in 1948 the National Party has pursued with single-mindedness of purpose the political ideology of apartheid through the amendment and supplementation of previous _ad hoc_ segregationist measures which had for a long time characterised South African economy, polity and society (Browett, 1982). To serve the interests of both Afrikanerdom and capitalism the apartheid state has not only preserved white supremacy, privilege and power but has also maintained the exploitation of Black labour as a lever of capital accumulation. To achieve these ends more systematic, comprehensive and strictly enforced control has been introduced over Blacks in their movement to urban areas and in their employment and residential opportunities. This has entailed, amongst other things the uprooting of Blacks from 'White areas' and their dumping in resettlement camps in Bantustans, the forced separation of families as a result of the tightening of influx control legislation, the bulldozing of Black inner city residential areas and urban squatter settlements, and the theft of citizenship rights through the masquerade of Bantustan 'independence'. Protest by workers, peasants and children against the racial oppression under which they are forced to live, work or study is met brutally and ruthlessly by an arsenal of state repression. In such circumstances South Africa affords illustration, from the dusty streets of monotonous urban townships to the destitute and impoverished Bantustans, of the imperative for geographers to confront 'unhappy' as well as 'happy' regions (Bunge, 1973).

Geographers based outside of South Africa have, in large part, failed to meet this challenge. As a result of this widespread scholarly neglect, the writings on South African social geography, unlike those on its history in which contributions from overseas-based academics are prominent, derive almost exclusively from the pens of a relatively small group of locally-based human geographers. From amongst these latter the academics teaching at the proliferating campuses of the Bantustans have, to date, effected little imprint so that the social geography literature is dominated by the research outpourings of the country's leading English- and Afrikaans- speaking universities. This work, in large part, reflects the deeply colonised nature of a discipline which has for long suffered from a cultural cringe complex brought on, as both cause and effect, by the meek acceptance of the academic imperialism emanating from the shores of Euro-America (Crush, Reitsma and Rogerson, 1982). Although the stirrings of an alternative homespun and decolonised human geography are beginning to surface at the present ime, the stamp of the borrowed changing trends and fashions in Anglo-American geography (sometimes in modified form) are firmly etched on the overall development of contemporary social geography in South Africa.

As one consequence of the diffusion of the various convolutions and revolutions in overseas geographic thought, no common or unified viewpoint which commands universal assent is to be found. Indeed, diversity and dissent would appear to be widespread, both between and amonst English- and Afrikaans- speaking geographers (Beavon, 1982a). At the polar extremes are those, on the one hand, who apologise for, legitimise or rationalise the social inequality of apartheid, whilst on the other, there are those who are committed to social science as a liberating force which continually seeks to expose the roots of exploitation and oppression. Others merely ponder whether the relationship of social geography to geography is analogous to that between social tennis and tennis.

From the midst of the tangled web of such disparate and often incompatible opinions and perspectives the task of this paper is to highlight, in an admittedly subjective, arbitrary and non-exhaustive fashion, the variable foci which comprise the past and present scope and purposes of social geography under apartheid. In this endeavour consideration will be given, where appropriate, to some of the related writings of social historians, social anthropologists, political scientists and sociologists. However, attention will be directed primarily to works by people who call themselves geographers, and more specifically to those which have been published from the late 1960s onwards - the period which marks the 'take-off' of social geography research in South Africa.

Strikingly, the term social geography, if mentioned at all, appears only fleetingly in previous sketches of the state of the art of South African geography (Beavon and Rogerson, 1981; McCarthy, 1982; Pirie and and Moon, 1982; Rogerson, 1982a; Wellings and McCarthy, 1983). Nevertheless, in the extant literature it is possible to differentiate four distinctive streams or traditions which, whilst not necessarily mutually exclusive, are substantially divergent. Of these the first two - (1) the works informed by the quantitative revolution, and (2) the analysis of the spatial dimensions of social problems, particularly the peculiar racial geometry of apartheid - do little more than mirror and replicate Anglo-American trends. In contrast, the latter two - (3) the literature rooted in the tradition of humanism and (4) and radical and/or neo-Marxist writings which seek a retheorisation of South African society in terms of class, capitalism and exploitation (Johnstone, 1982) - do attempt to make some potential innovatory contribution to an understanding of the conditions under which people are forced to live in South Africa. Each will be reviewed in turn.

Quantitative Social Geography

The move from descriptive analyses to those which employed statistical techniques and quantitative methods was warmly but tardiliy welcomed by South African geographers. Unfortunately, and as a result of their late arrival on the scene, the diffusion and acceptance of the quantitative revolution in South Africa coincided with a rising crescendo of criticism against its perspectives which was to ultimately result in its displacement from the centre stage of geography. Thus, whilst a growing questioning of the merits of quantitative approaches was increasingly evident in North America and Britain, these same approaches were securing a foothold and then a stranglehold on the direction of South African social geography research. Subsequently there has, to some extent, been an ebbing of the quantitative tide. This has been less so, though, in the social geography literature which emanates from Afrikaans-speaking geography departments. Indeed, the de-emphasis of quantitative approaches in the mass of English language contributions is said to lie at the root of the increasing divergence in the directions taken by English, as opposed to Afrikaans, geography in South Africa (Beavon, 1982a).

Whilst the scope of quantitative geography in South Africa was initially confined to delimiting the size, nature and classification of work and business places (Beavon and Rogerson, 1981), it rapidly found expression in social geography in the early 1970s through the application of quantitative procedures to the analysis of the spatial dimensions of the urban social fabric. One of the earliest studies which embraced the tradition of quantitatively examining the social geography of the city was Cohen and Hart's (1972) investigation of the residential clustering characteristics of a White minority population group in Johannesburg. This study, in common with much of the quantitative analyses which were to follow it, can be distinguished by its exclusive concern with White social space - a focus which was to be strengthened as the march of factor and other multi-variate analyses further bleached the local social geography. The pioneer applications of urban factorial ecology were confined entirely to investigations of the White segment of the apartheid city. In part, the thrust of this research was aimed at the spatial distribution of the White urban population because this permitted the results of these local analyses to be compared and contrasted either with patterns expected on the basis of the models and theories emanating from Anglo-American social geography or with the empirical results of research closely fashioned on the British and North American studies. In part also, the spatial bias towards the application of these procedures and models to the segregated White urban space can be attributed to the ready access to quantitative data for White urban area census tracts and to the non-availability or unreliability of data relating to the social milieu of the Black townships.

In general, White South African human geographers responded with alacrity to a move which was unwittingly regarded as one from the empirical-inductive method of Francis Bacon to the theoretical-deductive method of Karl Popper. It came to be recognised that new opportunities were opened up by the mathematisation of the analysis of the differentiation of spatial pattern and order in the residential mosaic of South African cities. Johannesburg was the locus for the initial studies into the socio-spatial structure of White urban South Africa (Hart, 1973, 1976a; Hart and Browett, 1976) and for the extension of the earlier work on the pattern of White minority groups in the city (Browett and Hart, 1977). More recently, however, there has appeared a continuing stream of bleached quantitative-inspired research into a variety of aspects of the White urban milieu in other parts of the country. Such studies include those on White residential mobility and neighbourhood satisfaction in Cape Town (Zietsman, 1980a, 1980b, 1981), the social elements of owner and rental housing in Cape Town (de Beer, 1978), the patterns of White elite residential development in Johannesburg (Hart, 1976a), social differentiation among White minority groups in Pretoria (van Bergen and Olivier, 1983), and the social impact of the initially White-focussed television service (Fairhurst, 1983). And at the same time there has also been a proliferation of replicative factorial ecologies of the White population residing in the country's major urban centres. Indeed, notwithstanding the appearance of a cautionary reflection on the difficulties encountered in applying factorial ecology to the multi-racial urban panorama of South Africa (Hart, 1978), the technique retains its appeal to the present day. Recent applications have extended its use (and abuse) to the socio-spatial dimensions of other race groups (Stickler, 1982), albeit not always successfully (Vlok, 1979).

Whilst such quantitative social geography may at times provide some interesting insights into a part of the local urban fabric, its contribution must be set against the enormity of South Africa's social problems, which it singularly fails to address. Theoretically moribund, wilfully neglectful both of the role of the state and of the structural-historical race- and class-based inequalities that shape the apartheid landscape, and implicitly or explicitly assuming a harmony of interests, or at least lack of open conflict, such studies do little to aid an understanding of the people and processes which lurk behind the patterns which are described in such sophisticated and elegant quantitative detail. This is indeed social geography as social tennis. For greater depth, empathy and sensitivity of analysis it is necessary therefore to turn attention to the three remaining traditions, each of which has sought to provide local social geography with a less Nero-like and hence more problem-oriented focus.

Colour it Darker - Geographers Study Social Problems

'Towards the close of the 1950s the massive urban exercise of social engineering through spatial manipulation, as implemented through the Group Areas acts, commanded the attention of a small group of South African researchers (Brookfield, 1957; Brookfield and Tatham, 1957; Kuper, Watts and Davies, 1958). Although submerged in the euphoria of quantification, the dissection of Durban's racial ecology by Kuper, Watts and Davies (1958) is now deservedly recognised as a classic extract in South African social geography. More than a decade was to pass, however, before the theme of social geography as focus for the analysis of social issues and inequality was to resurface in the South African geographical community, a rebirth which was associated not so much with the recognition of perceived inadequacies in the quantitative literature but rather with the diffusion of the liberal/relevance re-orientation of Anglo-American geography. A marked feature of this renewed commitment towards greater social responsibility and professional involvement was that the colour of social geography in South Africa progressively darkened in tone.

The first attempt to map the national geography of social well-being was contained in Board, Davies and Fair's (1970) portrayal of the space economy as a set of interrelated and interconnected nodes, networks and surfaces of economic development. The eviscerated economies of the Bantustans stood out starkly as the deepest poverty troughs in the spatial landscape. Upon the foundations established by this influential spatial organisation study there stemmed not only a clutch of ethnocentric diffusionist paradigm interpretations of socio-spatial inequalities (Browett, 1976, 1977; Browett and Fair, 1974; Schmidt, 1975) but also Smith's (1974) welfare geography analysis which highlighted the congruence of race and spatial inequalities in South Africa. The apartheid space economy was viewed as a variant of Friedmann's core-periphery model and of modernisation theory. The conceptual framework was that the polarisation of economy growth in the White core areas was accompanied by apartheid policies which contained or constrained the growth of urbanisation (Fair, 1972; Fair and Schmidt, 1974; Fair and Davies, 1976; Fair and Browett, 1979). At the same time, the enclaves of underdeveloping Black areas were regarded as essentially separate systems of spatial organisation which, although interspersed in location, were largely separated in function and imperfectly integrated with the innovative forces generated in the 'White' areas.

With the emergence of dependency paradigm and Marxist analyses of inequalities in South Africa the flawed interpretations within these socio-spatial inequality perspectives have been exposed. It has been recognised that, at best, their descriptions of the spatial organisation patterns gave undue emphasis to the juxtaposition of an incomplete body of information. They did not, and could not, explain the spatial differences because they did not take account of the underlying dynamics and forces of the over-arching political economy of which the patterns were a manifestation. For a long time, however, the contribution of geographers to such a reformulation, whilst potentially substantial, was minimal. Rather, the groundwork for class-based analyses of the exploitative forced labour economy was provided by studies from social historians, sociologists and political scientists. Only very recently have geographers begun to emerge from self-imposed exile and estrangement from their social science colleagues and in so doing have echoed this literature (Browett and Fair, 1979; Beavon and Rogerson, 1981; Crush, 1982; Fair, 1982; Crush and Rogerson, 1983).

In contrast, geographers have made a much more pronounced contribution at the urban scale of analysis. In an address to the South African Geographical Society in 1973 Davies (1974) chided his audience for their failure to confront such highly sensitive issues as the spatial restructuring of the country's cities in the face of the implementation of the group areas legislation. In response to this, in a series of recent studies by social geographers the extent and devastation wrought by the apartheid planner's demographic surgery are becoming manifest, as is the falsity of the government's claim that its involvement in urban population removals revolves around its concern to erase health hazards and slum conditions, particularly amongst the country's Coloured and Asian communities (Pirie, 1983a). In illuminating the enactment, implementation, outcome and reception or urban racial segregation, Pirie (1984a) shows that while the state poses as an agent of slum upgrading and moderator of racial conflict, its actions reveal naked disruption, opportunism and the politicalisation of law and rights in South Africa. Likewise, in studying the ghettoisation of Cape Town by residential apartheid, Western (1982) reveals the underlying motivation for the whole programme of urban removals as part of the geography of urban social control. Perceptively, Western (1982, 218) notes that "the White minority has been trying to maintain its rule by segregating its darker, outnumbering and poorer fellow-citizens. The social geography of South African cities is a product of witting intent; the hand is not hidden here".

At the same time, the making of the segregated Black townships and efforts to understand their unique social fabric have emerged as new research frontiers in the de-whitening and decolonising of South African social geography studies. In crossing into the hitherto unknown world of the townships the research interests of social geographers began to converge with those of other social scientists writing 'popular' or working class history (see Bozzoli, 1979, 1983a). The crystallisation of an important literature, styled as the new wave of African historiography, is extensively reviewed elsewhere (Crush and Rogerson, 1983). Suffice it here, therefore, to merely note the geographical contributions to understanding the creation of townships (Beavon, 1982b), the excision of Blacks from former leasehold areas in White urban space (Pirie and Hart, 1984), the historical struggles for survival in the informal sector (Beavon and Rogerson, 1980; Rogerson, 1983b; Rogerson and Beavon, 1984; Rogerson and Hart, 1985) and urban township bus boycotts (Pirie, 1983b).

Contemporary social geographic investigation of townships have been predominantly focussed on issues of shelter and poverty. In terms of accommodating Blacks in the city, geographers have probed the potential of self-help housing schemes (Hart and Hardie, 1983) illuminated the allocation procedures of township housing (Mashile and Pirie, 1977) and exposed the barrack-like hostels as serving the interests of state and capital (da Silva, 1983). The poverty studies have revolved around the extensive investigation of life and survival in the contemporary informal sector of South African cities (Rogerson and Beavon, 1980, 1982; Beavon and Rogerson, 1982, 1984; Davidson, 1984; Tomaselli, 1984). Indeed the volume of ongoing research on the urban informal economy of South Africa suggests that it today commands a central place on the research agenda of the country's socially concerned geographers.

Outside of the urban realm few geographers have investigated the social conditions of the Bantustans. To date the most notable works have emerged from Letsoalo's (1982) study of rural survival strategies in the Lebowa Bantustan. The rural tragedy of forced mass population upheavals and dumping in the Bantustans have, however, been documented by non-geographers (Desmond, 1971; Mare, 1980; Marks, 1980). From the pens of geographers have appeared contributions exposing the desperate conditions of life in the areas of rural resettlment (Rogerson and Letsoalo, 1981; Letsoalo, 1982) known as closer-settlements. In addition, further study revealed that rural lifestyles were little better under the programmes of rural improvement (Letsoalo and Rogerson, 1982). Exposing the bleakness of daily life in the Bantustans as a counter to the government painted pictures of slumbering rural retreats has thus emerged as a further local point for social geographic endeavour.

While the concern for inequalities has re-oriented relevant social geography research to the _terra incognita_ of the world of townships and the Bantustans, it remains to be acknowledged that some notably problem-focussed on White social space have also been produced. As before, the research agenda relating to social problems of White South Africans largely parallels work undertaken in Europe or North America. For example, during the 1970s the gaze of local geographers fell upon the spatial dimensions of crime (Nisbet, 1974; Ward, 1978), health care (Pirie, 1976) and the delivery of public services (Winckler, 1978). In the addition, an insight was secured into the special problems of the White elderly (Van der Merwe, 1979), those with sensory handicaps (Van Zyl, 1979) and those suffering from mental illness (Van der Merwe and Zietsman, 1982). Interest continues to be expressed in the problems of White minority groups as evidenced in research on Greek (Spanoudes, 1982) and Mauritian (Martin, 1983) immigrant communities. The housing difficulties of Whites and the ameliorative roles of private (Martin, 1982) and public sector agencies (Pirie, 1982a; Parnell, 1983) have spawned some innovative research. Indeed, that 'unhappy' regions exist in White as well as Black South Africa has been clearly exposed in research flowing from Bunge-style urban expeditions mounted annually in Johannesburg since 1978 (Mabin, 1984). Although social geographers recorded the existence of Blacks living in White urban space (Hart, 1976b), the important task of penetrating the silent world of exploitation daily experienced by domestic servants was left to social anthropologists (Preston-White, 1982) and sociologists (Cock, 1980).

In summary, it can be seen that the redirection of social geographic concern in favour of a problem-oriented focus has dramatically altered the directions of research. The Soweto riots of 1976 forcibly brought home the fact that geographers in South Africa had nothing to say about the conditions of life facing the mass of their country's population. In a painful process of readjustment geographers are now beginning to confront full square the unsavoury and unsocial dimensions of living under apartheid. Consequently, the social geography of South Africa is increasingly darkening in hue. This trend is further accentuated by contributions from two newer traditions in South African geography - those involving the adoption of either humanist or radical/neo-Marxist approaches in research.

The Humanist Infusion

Humanist writings point to and seek to correct for geographers' meagre understanding of the ways in which places are perceived and experienced by their inhabitants. Over the past five years the advent of such humanist-inspired research has added considerable depth to observations across the widening breadth of the South African social geography terrain. It is notable here that the initial humanist enquiry relating to South Africa was pursued by a non-South African based geographer. In terms of 'knowing one's place', Western (1978, 1981) proved the experiential world of Cape Town's Coloured community. Through essaying the texture of its inhabitants' lives, Western uncovered the quintessential meaning of place, mapping everyday geographies of distress, disadvantage, status and fear. The cruel inhumanity of apartheid was exposed naked in the bleak experiences of the Mowbray Coloured community as it was forced to uproot under the exigencies of group areas 'planning'.

Exploration of the world of the Black townships by White South African researchers is more constrained than the areas of coloured settlement in Cape Town where Western worked. Recognising that some aspects of reality can only be known through direct experience and that White academics are denied the gaining of this by being forbidden to live in townships, some geographers turned instead to examine creative literature as source material. The first local foray in using literature as a research tool was confined to highlighting the simultaneous anomie and exuberance of night-time city life experienced in all capitalist societies (Butler-Adam, 1981). It was left to Pirie (1982b) to demonstrate the potential richness of creative writings as a vehicle for exploring the existential realities of life in Johannesburg, especially in the Black township areas. The task of realising potential has been successfully undertaken by Hart (1982, 1984a, 1984b). In what must rank as a major contribution to local social geography, Hart and Pirie (1984) evoke the lost spirit of Sophiatown, a former area of Black settlement in inner Johannesburg which was demolished in the later 1950s. Against the objective slum characteristics of over-crowding, squalor, violence and poverty are balanced the subjective personality of the place, the endearing attributes of community, freedom, vitality and variety which are celebrated in Sophiatown literature. In a further study Hart (1984b) contrasts the subjective world of Sophiatown, a more or less spontaneous settlement in its origins, with Soweto, progeny of apartheid population removals. Sharply revealed is the marked divide between the observable facets of places as emphasised by officialdom and apartheid apologists - the racial purity and neat orderliness of Soweto and the rundown slumlike appearance of Sophiatown - and the qualitative characteristics of places as expressed in creative literature - the emotional affinity of residents with Sophiatown and their emotional alienation from Soweto. Literary geographical studies are thus enriching greatly an appreciation of the unsocial world daily endured by Black South Africans.

In an interesting contribution from Afrikaans geography, Steenkamp (1983) maps the landscapes of fiction contained in the prose of C.J. Langenhoven. The emphasis here upon the White rural rather than the Black urban setting situates this work in the vein of mainstream literary geographical studies as pursued in North America and Western Europe. Reasserting the local South African scene as a source for innovation, Hart and Rogerson (1985) employ creative literature to illumine the hidden experiential world of those struggling for existence in the informal sector. This examination of shebeening through the medium of English literature (Hart and

Rogerson, 1985) once more reinforces the potential for humanist social geography research in South Africa to expand the horizons, newly opened by other research methods. Finally, and also breaking fresh ground, is Pirie's (1984b) elucidation of the process of naming Soweto. The experience of Soweto conclusively reveals that place-naming is not only the outcome of private and impenetrable contemplation but can also be a social activity.

Radical Perspectives Awaken

The awakening of a radical stream in South African social geographic enquiry owes its strength more to local considerations than to the diffusion of the North American radical/Marxist geography revolution. The roots of radical social geography in South Africa are essentially threefold. First, there is the radicalising influence upon social science in general of recent political events, particularly the Soweto uprising in 1976 and the successful decolonisation, under socialist leadership, of Mozambique, Angola and Zimbabwe (Beavon and Rogerson, 1981). Second, at certain English-speaking South African universities which at least tolerate radical thinking there is a long tradition of social criticism, "... even if it prefers to cultivate liberal criticism aimed at incremental reform and/or the deracialization of South African capitalism" (McCarthy, 1982, 58). Finally there is the appearance in the 1970s of a new school in South African studies which, in one way or another, can be distinguished by the adoption of historical materialism and class analysis as its central focus (Johnstone, 1982).

With one exception, the themes represented in the extant radical social geography writings on South Africa mirror those under scrutiny in the liberal-welfare and humanist interpretations. Many radical writers, however, aim to balance concern for explicating the economic basis of domination with some involvement in political praxis (McCarthy, 1982; Wellings and McCarthy, 1983). In the radical literature the main emphasis is upon the political economy of urbanism, shelter and poverty, particularly as these themes relate to townships. Recent critical contributions from geographers have dissected the state's post-1976 'reformist' initiatives in meeting crises in the spheres of employment and housing for urban Blacks. On shelter, Lea (1982) avers that the government's relaxation of controls on housing tenure and its growing enthusiasm for self-help initiatives must be seen as primarily functional to the interests of capital. This conclusion is reinforced by important submissions by Mabin (1983) and Mabin and Parnell (1983) on the political implications for Black South Africans of the state's recent shift to a policy of selling off previously-rented township housing. Equally gloomy are the findings of the informal sector studies which are informed by radical theory. These suggest that there is little prospect of ameliorating township poverty through relaxing controls and promoting street trading, petty production and services (Beavon and Rogerson, 1982; Rogerson and Beavon, 1982; Tomaselli, 1984).

Prominent research thrusts are also to be found in retheorisation and steps towards a class analysis of the local 'urban questions' rooted in conflict around the built environment (McCarthy, 1981; McCarthy and Friedman, 1981, 1983; McCarthy and Smith, 1981). The involvement of geographers in local community struggles has dovetailed their research with that of historians and political scientists analysing working class resistance to urban renewal, squatter movements, rent struggles and transportation boycotts (Proctor, 1979; Stadler, 1979; Lodge, 1981, 1983). Indeed, as geographers begin to address the ahistorical nature of South African social geography studies so there is a convergence, particularly with the procedures of radical social historians such as Van Onselen. In terms of the changing foci and directions of geographical enquiry in South Africa the works of Van Onselen (1982) are now recognised as outstanding submissions, informed by historical materialism, on people's everyday social experience and struggles.

The one new focus of radical social geography concerns the hitherto neglected world of women's studies. Against a backdrop of a surge in feminist interpretations in South African studies (Bozzoli, 1983b) as a whole, it is perhaps not surprising to note that the dawn of geographical research in this direction. Within the current literature are contained only two studies by geographers. These examine, respectively, the changing participation of women in industry (Martin and Rogerson, 1984) and of women's role in the South African informal sector (Beavon and Rogerson, 1984). However, a groundswell of feminist research is building up, particularly in the English-speaking universities, so that further South African contributions towards a radical social geography of women may be anticipated in the next few years.

Conclusions

Over the last decade and a half South African social geography has undergone substantial reorientations in direction and scope, changes which are regarded by many as comprising, if not titanic strides forward, then at least progress. Works which merely gave quantitative expression to the nature and form of spatial differentials in social conditions are now seen as comprising a conceptually inadequate and defective framework for analysis, as are those which, implicitly at least, merely sought a more morally-justifiable and tolerable capitalist order. Both perspectives have been displaced from centre stage by those which attempt to understand how society is organised in space either through the aegis of humanist traditions or through the examination of the uneven march of capitalist social relations of production over the landscape. In these endeavours the focus of attention is fixed firmly on the unsocial world of apartheid, particularly as it is experienced by Blacks in the Bantustans and the townships. While still in the embryonic stages, the perspectives encompassed by these studies do entail a progressive loosening of the previously-dominant and emasculating ties to Anglo-American human geography. With the rejection of South Africa as a dumping ground for expatriate theories and methods of analysis there has thus arisen the hope that the study of South African society by social geographers can constitute a potential source for innovation and advance.

If, in the future, a de-whitened local social geography were to emerge as the dominant perspective then it is envisaged that it could well make some positive impact, however humble, to the geography and social science literature both in South Africa and overseas. And in so doing, of course, reinforce the rising body of opinion and thought which is presently flowing to the cores from the periphery. The likelihood of such an eventuality does, on three counts at least, appear promising. Firstly, social geography should greatly benefit from its links with social history, the discipline which will very probably be in the vanguard of social science research in the 1980s and one which already has a solid and distinguished presence in South African studies. The challenge here is for social geography to break its previous habit of parasitic borrowings, not merely to hang on to the coat-tails of the social historians but to supplement and extend their work.

Secondly, the directions taken in the social geography studies of shelter, poverty and inequality can, and should, have much wider significance. South Africa may not only be the polecat of the world but also a microcosm of it in which the central issues to be addressed stand out more forcefully and with greater poignancy. In part, success in meeting this challenge is dependent on the production of book-length sustained research in which greater emphasis is placed on comparative analysis approaches. Thirdly, significant contributions should emanate from social geography analysis of the role of the state and its use of space as an explicit and crucial instrument of public policy. Specifically, the examination of the race-space-class conjuncture in South Africa affords a fertile ground in which to search for the mediating concepts which are thought to exist between agency and structure. In this manner contemporary social geographers, unlike their predecessors, may assist in the derivation and articulation of social theory.

Yet doubts still remain as to the significance both of social geography and of academic social science research of which it forms a part. Whilst these reservations are not peculiar, or confined, to South Africa, they are more acutely felt and most starkly revealed in that context. In the first instance, what must be clearly enunciated, is the value and distinctiveness of the contribution that social geography can make in enriching our understanding of society, in South Africa and elsewhere. Do, in fact, social geographers have an esteemed and worthwhile role to play, and if so what is it? Certainly, human geographers have long claimed that something can be done to improve the standards of material well-being of impoverished people and that their discipline is well suited to provide some academic contribution through a sympathetic and compassionate analysis of problems experienced by poor people. Others, however, have remarked that "the work of geographers is a production without consumers" (Quaini, 1982, 169). What needs to be confronted here, especially in a situation in which constraints are placed on the opportunities for field work, is the explication of the specific purpose of social geography.

At the same time, and in more general terms, what also has to be faced is the nature of the relationship of the social geographer and social scientist to society. This poses considerable dilemmas which, whilst again universal, are most brutally and overtly exposed in South Africa. Specifically, it is difficult to envisage how the writing of treatises on what has been, can be or must be done will lead to reductions in the oppression of Black people or to their liberation and self-emancipation. More cynically, the burgeoning social geography literature can be regarded as meeting the internal needs of those involved in the writing of it, a part of an ivory tower game in which achievement is measured by the gaining of academic kudos and the expansion of *curriculum vitae*. Less cynically, though, what may be questioned is the utility of devoting additional research resources to demonstrating over and over again, but from different perspectives, the pauperisation and suffering of Black people. In these circumstances it can be argued that there is need not for more understanding of inhumanity, oppression and exploitation but of a need to end it. Working towards this goal in South Africa does, however, present social geographers and social scientists with some difficulty, particularly when little faith can be placed in the efficacy of formulating and advocating local reactive, ameliorative and incrementalist policies designed to facilitate progressive social change. Despite trust in the growth prospects of little acorns, such prescriptions at best suffer from an unfavourable comparison with attempts to re-arrange deck chairs on the Titanic, and at worst serve mainly to prop up and preserve the established social, economic and political order. If, with Engels, it is believed that it is folly to hope for an isolated settlement of individual social issues, then the alternative prospect is that of a radical overthrow of the status quo. And in South Africa this is indeed the case for the inevitable, but not predictable, outcome of increasing repression and authoritarian control is that of violent revolutionary change. In the face of this future and the place of social geography studies and the role of social geographers in South Africa becomes ever more problematic.

REFERENCES

Beavon, K.S.O. 1982a Die smouse van Johannesburg: lesse vir menslike geografie. P. Serton Memorial Lecture, University of the Orange Free State,

Bloemfontein, October

Beavon, K.S.O. 1982b Black townships in South Africa: terra incognita for urban geographers, South African Geographical Journal 64, 1-20

Beavon, K.S.O. & C.M. Rogerson 1980 The persistence of the casual poor in Johannesburg, Contree 7, 15-21

Beavon, K.S.O. & C.M.Rogerson 1981 Trekking on: recent trends in the human geography of southern Africa, Progress in Human Geography 5, 159-189

Beavon, K.S.O. & C.M. Rogerson 1982 The 'informal sector' of the apartheid city: the pavement people of Johannesburg, in D.M. Smith (ed.), Living Under Apartheid, London, George Allen & Unwin, 106-123

Beavon, K.S.O. & C.M. Rogerson 1984 The changing role of women in the urban informal sector of Johannesburg, South Africa. Paper presented to the IGU Working Group on Urbanisation in Developing Countries Symposium, Lille, August

Board, C., R.J. Davies & T.J.D. Fair 1970 The structure of the South African space economy: an integrated approach, Regional Studies 4, 367-92

Bozzoli, B. (ed) 1979 Labour, Townships and Protest: Studies in the Social History of the Witwatersrand, Johannesburg, Ravan

Bozzoli, B. (ed) 1983a Town and Countryside in the Transvaal: Capitalist Penetration and Popular Response, Johannesburg, Ravan

Bozzoli, B. 1983b Marxism, feminism and

		South African studies, Journal of Southern African Studies 9, 139-171
Brookfield, H.C.	1957	Some geographical implications of the apartheid and partnership policies in southern Africa, Transactions of the Institute of British Geographers 23, 225-47
Brookfield, H.C. & M.A. Tatham	1957	The distribution of racial groups in Durban, Geographical Review 47, 44-65
Browett, J.G.	1976	The application of a spatial model to South Africa's development regions, South African Geographical Journal, 58, 118-129
Browett, J.G.	1977	Export base theory and the evolution of the South African space economy, South African Geography Journal, 59, 18-29
Browett, J.G.	1982	The evolution of unequal development within South Africa: an overview, in D.M. Smith (ed) Living under Apartheid: Aspects of Urbanization and Social Change in South Africa, London, Allen and Unwin, 10-23
Browett, J.G. & T.D.J. Fair	1974	South Africa 1870 to 1970: a view of the spatial system, South African Geographical Journal, 56, 111-120
Browett, J.G. & T. Hart	1977	The distribution of White minority groups in Johannesburg, South African Geographer, 404-421
Bunge, W.	1973	Exploration in Guadeloupe: region of the future, Antipode

		5(2), 67-83
Butler-Adam, J.F.	1981	Literature and the night time geography of cities, South African Geographical Journal 63, 47-59
Cock, J.	1980	Maids and Madams: A Study in the Politics of Exploitation, Johannesburg, Ravan
Cohen, S. & G.H.T. Hart	1972	Clustering characteristics of a migrant community in the core of the primate centre, South African Geographical Journal 54, 85-94
Crush, J.S.	1982	The Southern African regional formation, Tijdschrift voor Economische en Sociale Geografie 73, 200-12
Crush, J. & C.M. Rogerson	1983	New wave African historiography and African historical geography, Progress in Human Geography 7, 203-231
Crush, J., H. Reitsma & C.M. Rogerson	1982	Decolonizing the human geography of Southern Africa, Tijdschrift voor Economische en Sociale Geografie 73, 197-198
Da Silva, M.	1983	Black hostels and compounds in Johannesburg and Soweto; their characteristics and significance. Unpublished paper, Department of Geography and Environmental Studies, University of the Witwatersrand, Johannesburg
Davidson, J.	1984	Hawking in Finge Village, Grahamstown - a working paper. Paper presented to the 2nd Carnegie Inquiry into poverty and development in Southern Africa,

Cape Town, April

Davies, R. 1974 Geography and society, South African Geographical Journal 56, 3-14

De Beer, A.S.A. 1978 Verskille in die kenmerke van eienaars en huurders in die Suidelike Voorstede van Kaapstad, Geo-Stell 2, 1-11

Desmond, C. 1971 The Discarded people, Harmondsworth, Penguin

Fair, T.J.D. 1972 The Metropolitan Imperative, Inaugural lecture, Witswatersrand University Press, Johannesburg

Fair, T.J.D. 1976 Polarisation, dispersion and decentralisation in the South African space economy, South African Geographical Journal, 58, 40-56

Fair, T.J.D. 1982 South Africa; Spatial Frameworks for Development, Cape Town, Juta

Fair, T.J.D. & J.G. Browett 1979 The urbanisation process in South Africa, in D.T. Herbert and R.J. Johnston (eds) Geography and the Urban Environment: Progress in Research and Applications, Vol. 2, New York, Wiley, 259-94

Fair, T.J.D. & R.J. Davies 1976 Constrained urbanisation; white South Africa and Black Africa compared, in B.J.L. Berry (ed) Urbanisation and Counterurbanisation, Beverly Hills, Sage, 145-168

Fair, T.J.D. & C.F. Schmidt 1974 Contained urbanisation: a case study, South African Geographical Journal 56, 155-66

Fairhurst, U.J. 1983 The spatial diffusion of television receivers and the social impact of this innovation in the Randburg area, M.A. thesis, University of South Africa

Hart, D.M. 1982 Johannesburg's Black urban townships in English literature. Unpublished paper, Department of Geography and Environmental Studies, University of the Witwatersrand, Johannesburg

Hart, D.M. 1984a South African literature and urban Black townships, M.A. dissertation, University of the Witwatersrand, Johannesburg

Hart, D.M. 1984b Place and placelessness in two Black South African townships. Paper presented at the Association of American Geographers Annual Conference, Washington D.C., April

Hart, D.M. & G.H. Pirie 1984 The sight and soul of Sophiatown slum, Geographical Review 74, 38-47

Hart, D.M. & C.M. Rogerson 1984 Literary geography and the informal sector, Geography Research Forum 8

Hart, T. 1973 An examination of sectoral and concentric patterns in the socio-economic structure of Johannesburg, South African Geographical Journal 55, 32-39

Hart, T. 1976a The evolving pattern of elite White residential areas in Johannesburg, 1911-1970, South African Geographical Journal 58, 68-75

Hart, T. 1976b Patterns of Black residence in the White residential areas of

Johhannesburg, _South African Geographical Journal_ 58, 141-50

Hart, T. 1978 Factorial ecology and the South African city; some problems and prospects, _South African Geographer_ 6, 123-129

Hart, T. & J.G. Browett 1976 _A Multivariate Spatial Analysis of the Socio-Economic Structure of Johannesburg_, Johannesburg, University of the Witwatersrand, Urban and Regional Research Unit, Occasional Paper No. 13

Hart, T. & G.J. Hardie 1983 _Factors Motivating Residential Consolidation in Selected South African Self-Help Housing Contexts_, Johannesburg, National Institute for Personnel Research CSIR Special Report PERS 358

Johnstone, F. 1982 "Most painful to our hearts": South Africa through the eyes of the new school, _Canadian Journal of African Studies_ 16, 5-26

Kuper, L., H. Watts & R.J. Davies 1958 _Durban: A Study in Racial Ecology_, London, Jonathan Cape

Lea, J.P. 1982 Government dispensation, capitalist imperative or liberal philanthropy? Responses to the Black housing crisis in South Africa, in D.M. Smith (ed) _Living Under Apartheid_, London, George Allen & Unwin, 198-216

Letsalo, E.M. 1982 Survival strategies in rural Lebowa: a study in the geography of poverty, M.A. dissertation, University of the Witwatersrand, Johannesburg

Letsoalo, E.M. & C.M. Rogerson 1982 Rural 'development' under apartheid: betterment planning in Lebowa, South Africa, *Geoforum* 13, 301-314

Lodge, T. 1981 The destruction of Sophiatown, *Journal of Modern African Studies* 19, 107-32

Lodge, T. 1983 *Black Politics in South Africa Since 1945*, Johannesburg, Ravan

Mabin, A.S. 1983 'State homes for sale': re-examining home ownership, *Work in Progress* 26, 4-7

Mabin, A.S. 1984 WEJGE; the genesis of an exploration in geographical learning, *South African Geographer* 12, 69-76

Mabin, A.S. & S. Parnell 1983 Recommodification and working-class home ownership: new directions for South Africa cities?, *South African Geographical Journal* 65, 148-166

McCarthy, J.J. 1981 Research on neighbourhood activism: review, critique and alternatives, *South African Geographical Journal* 63, 107-131

McCarthy, J.J. 1982 Radical geography, mainstream geography and Southern Africa, *Social Dynamics* 8, 53-70

McCarthy, J.J. & M. Friedman 1981 Black housing, ideology and the media in South Africa, 1970-1979, *Critical Arts* 2(2), 51-66

McCarthy, J.J. & M. Friedman 1983 On the social construction of urban problems, in K. Tomamselli and R. Tomaselli (eds),

		Culture, Ideology and the Media in South Africa, Johannesburg, Jonathan Ball
McCarthy, J.J. & D.P. Smith	1981	The political economy of urban land use and land use planning: review of the theory with conclusions for South Africa, Social Dynamics 7, 8-31
Mare, G.	1980	African Population Relocation in South Africa, Johannesburg, South African Institute of Race Relations
Marks, S.	1980	From Difaqane to discarded people: South Africa's internal refugees, African Research and Documentation 22, 2-12
Martin, V.	1982	Housing White miners in South Africa: a preliminary study. Unpublished Paper, Department of Geography and Environmental Studies, University of the Witwatersrand, Johannesburg
Martin, V.	1983	Mauritians in South Africa: a study in population geography. Unpublished paper, Department of Geography and Environmental Studies, University of the Witwatersrand, Johannesburg
Martin, V. & C.M. Rogerson	1984	Women and industrial change: the South African experience, South African Geographical Journal 66, 32-46
Mashile, G.G. & G.H. Pirie	1977	Aspects of housing allocation in Soweto, South African Geographical Journal 59, 139-149

Nisbet, B.G.	1974	Aspects of crime in the central area of Durban, _Isizwe_ 1, 38-51
Parnell, S.	1983	Nature and development of White worker housing in Johannesburg. Unpublished paper, Department of Geography and Environmental Studies, University of the Witwatersrand, Johannesburg
Pirie, G.H.	1976	Apartheid, health and social services in Greater Johannesburg, in D.M. Smith (ed), _Separation in South Africa: Homelands and Cities_, London, Queen Mary College, Occasional Paper No. 7, 85-97
Pirie, G.H.	1982a	Sleepers beside tracks: housing in South Africa's state railway corporation, _South African Geographical Journal_ 64, 144-154
Pirie, G.H.	1982b	Mostly 'Jubek': urbanism in some South African English literature, _South African Geographical Journal_ 64, 63-71
Pirie, G.H.	1983a	Urban population removals in South Africa, _Geography_ 68, 347-349
Pirie, G.H.	1983b	Urban bus boycott in Alexandra township, 1957, _African Studies_ 42, 67-77
Pirie, G.H.	1984a	Race zoning in South Africa: board, court, parliament, public, _Political Geography Quarterly_ 3, 207-21
Pirie, G.H.	1984b	Letters, words, worlds: the naming of Soweto, _Africa Studies_ 43, 43-51

Pirie, G.H. & D.M. Hart	1984	The urban process in Johannesburg's Western Areas, 1905-1962, Journal of Urban History
Pirie, G.H. & B.P. Moon	1982	Six decades of research for higher degrees in geography in South Africa, South African Geographer 10, 197-202
Preston-Whyte, E.	1982	Segregation and interpersonal relationships: a case study of domestic service in Durban, in D.M. Smith (ed) Living Under Apartehid, London, George Allen & Unwin, 164-182
Proctor, A.	1979	Class struggle, segregation and the city: a history of Sophiatown, 1905-40, in B. Bozzoli (ed) Labour Townships and Protest, Johannesburg, Ravan, 49-89
Quaini, M.	1982	Geography and Marxism, Oxford, Blackwell
Rogerson, C.M.	1983a	The citation of scientific serials in South African geographical journals, 1975-1980, Humanitas 9, 251-254
Rogerson, C.M.	1983b	The casual poor of Johannesburg, South Africa: the rise and fall of coffee-cart trading. Ph.D. dissertation, Queen's University, Kingston, Ontario.
Rogerson, C.M. & K.S.O. Beavon	1980	The awakening of 'informal sector' studies in southern Africa, South African Geographical Journal 62, 175-190
Rogerson, C.M. & K.S.O. Beavon	1982	Getting by in the 'informal sector' of Soweto, Tijdschrift voor Economische en

Sociale Geografie, 73, 250-265

Rogerson, C.M. & K.S.O. Beavon 1984 A tradition of repression: the street traders of Johannesburg, in R. Bromley (ed) Planning for Small Enterprises in Third World Cities, Oxford, Pergamon

Rogerson, D.M. & D.M. Hart 1984 The survival of the 'informal sector': the shebeens of Black Johannesburg, South African Geographical Journal 67

Rogerson, C.M. & E.M. Letsoalo 1981 Rural underdevelopment, poverty and apartheid: the closer settlements of Lebowa, South Africa, Tijdschrift voor Economische en Sociale Geografie 72, 347-361

Schmidt, C.F. 1975 A spatial model of authority-dependency relations in South Africa, Journal of Modern African Studies 13, 483-490

Smith, D.M. 1974 Race-space inequality in South Africa: a study in welfare geography, Antipode 6, 42-69

Spanoudes, R.N. 1982 The shop down the road: a preliminary study of Greek cafes in South Africa, B.A. Hons. dissertation, University of the Witwatersrand, Johannesburg

Stadler, A. 1979 Birds in the cornfields: squatter movements in Johannesburg 1944-47, in B. Bozzoli (ed) Labour, Townships and Protest, Johannesburg, Ravan, 19-48

Steenkamp, W.J.	1983	Die wêreld van C.J. Langenhoven, Geo-Stell 7, 1-8
Stickler, L.	1982	Aspects of the socio-economic and spatial organisation of the Coloured population on the Witwatersrand, M.A. dissertation, University of the Witwatersrand, Johannesburg
Tomaselli, R.	1984	The Indian Flower Sellers of Johannesburg, M.A. dissertation, University of the Witwatersrand, Johannesburg
Van Bergen, M.M. & J.J. Olivier	1983	Sosiale differensiasie tussen Blanke etniese groepe in Pretoria, South African Geographer 11, 41-48
Van de Merwe, I.J.	1979	Residensiele vestigingspatrone van Blanke bejaardes in Kaapstad, South African Geographer 7, 11-20
Van der Merwe, I.J. & H.L. Zeitsman	1982	Ruimtelike patrone van geestessiektes in Kaapstad, South AFrican Geographer 10, 9-18
Van Onselen, C.	1982	Studies in the Social and Economic History of the Witwatersrand 1886-1914, 2 vols., Johannesburg, Ravan
Van Zyl, G.S.	1979	Stedelike aktiwiteitspatrone en omgewingskennis van sintuiglik-gestremde skoliere in Worcester, Geo-Stell 3, 54-64
Vlok, A.C.	1979	'n vergelykende faktor-ekologiese studie van die verskillende bevolkingsgroepe in Kaapstad, Geo-Stell 3, 1-11

Ward, G.J.	1978	Spatial autocorrelation and the analysis of patterns resulting from crime occurrence, M.A. dissertation, Rhodes University
Wellings, P.A. & J.J. McCarthy	1983	Whither southern African human geography? Area 15, 337-345
Western, J.	1978	Knowing one's place: 'the Coloured people' and the Group Areas Act in Cape Town, in D. Ley and M.S. Samuels (ed), Humanistic Geography: Prospects and Problems, London, Croom Helm, 297-318
Western, J.	1981	Outcast Cape Town, London, George Allen & Unwin
Western, J.	1982	The geography of urban social control: Group Areas and the 1976 and 1980 civil unrest in Cape Town, in D.M. Smith (ed) Living Under Apartheid, London, George Allen & Unwin, 217-229
Winckler, T.	1978	Die beskikbaarheid van openbare diensfasiliteite binne die sosio-ekonomiese statussones van die Kaapstadse munisipale gebied, Geo-Stell 2, 40-49
Zeitsman, H.L.	1980a	Reuimtelike patrone van Blanke residensiele mobiliteit in die Kaapstadse metropolitaanse gebied, South African Geographer 8, 113-125
Zeitsman, H.L.	1980	'n toepassing van die Markov-model op residensiele mobiliteit in die Kaapse metroplitaanse begied, South African Geographical

		Journal 62, 18-32
Zietsman, H.L.	1981	Woonbuurttevredenheid in die Kaapse metropolitaanse gebied, South African Geographical Journal 63, 132-146

Chapter Nine

SOCIAL GEOGRAPHY AND DEVELOPMENT IN SUB-SAHARAN AFRICA

Joe Doherty

An examination of social geography in the context of Sub-Saharan Africa raises two immediate definitional issues: what is 'social geography' and what is 'Sub-Saharan Africa'? The latter is more easily dealt with. As employed here, Sub-Saharan refers to that territory comprising 39 countries and some 350 million people, which is bounded in the north by the Sahara desert and in the south by the borders of the Republic of South Africa. The region embraces a great diverstiy of ecological conditions and forms of social, political and economic organisation. The diversity of organisation was, however, greater in the past than at present. The homogenising forces of European colonialism and international capitalism have by turns obscured, irrevocably changed or obliterated many indigenous structures. Yet the tenacity of the 'traditional', in its patchy survival and in its capacity to adapt, ensures that diversity remains as much characteristic of the region as any uniformities imposed by external influences.

A definition of social geography in the Sub-Saharan context is more problematic. For while we can talk about the social geography <u>of</u> Sub-Saharan Africa, the formulation of a definition of social geography <u>in</u> Sub-Saharan Africa is less easily achieved. A brief consideration of an early attempt by Gilbert and Steel (1945) to define the concerns of social geography in a colonial and African context exemplifies the problem. They identified four main branches: the distribution of population, the distribution and form of rural settlement, the geography of towns and cities and the distribution of social groups and their 'way of life' in different environments. As a scenario for African social geography this definition is deficient in two important respects. First, it is an atheoretical definition concerned only with the identification of subject matter, with empirical context. It gives little or no consideration to approaches and methodology beyond what can be implied from the statement that 'maps' are 'the principal tools' of the social geographer (p. 118). Indeed Gilbert and Steel's is a residual definition, they equate social geography to human geography with the economic (and political?) elements removed; it is a diffuse definition with no clear boundaries or identifiable theoretical core. Second, the definition is a derived definition, formulated elsewhere, in the entirely different social and economic conditions of Britain, and transferred to colonial Africa. Beyond their own personal commitment, the motivation of Gilbert and Steel in making this transfer was to meet the ostensible aims and objectives of the 1940 Colonial Development and Welfare Act which identified 'social welfare' along with 'economic progress' as crucial ingredients of colonial development programmes. Additionally Gilbert and Steel were motivated by a desire to enhance the status of geography which had been noticeably ignored or neglected in several important African research reports of the period (e.g. Hailey, 1938).

Despite these deficiencies, there appear to have been no serious attempts by other authors to identify alternative directions for the field. Indeed on the rare occasions when definitions have been offered they have reinforced rather than challenged that of Gilbert and Steel (Kay, 1967; Prothero, 1972). Consequently social geography in Africa has had no clear self-image, it has lacked a consciousness of its own existence as an identifiable indigenous academic specialism. Its content and theoretical priorities have been shaped largely by exogenous influences emanating from the European, and latterly, North American heart-lands of academic investigation. The extent to which social geography as practised in Africa has been and continues to be dominated by 'Africanists' (alien experts in African affairs) is symptomatic; even the growing body of indigenous African scholars now studying and researching in the field display, with few exceptions, the imbibed values of alien - to be explicit, Western - education. As recent overviews suggest, the principal contribution of sub-Saharan studies to the general development of social geography, has been as empirical support for theoretical propositions determined elsewhere (Jones, 1975; Jones and Eyles,

1977; Jackson and Smith, 1984). Cultural imperialism is as prominent in Africa's intellectual landscape as economic imperialism in the physical landscape.

These issues are not raised necessarily to condemn, though some condemnation may be involved. Rather, their identification is seen as a necessary prelude to an examination of the changing nature of social geography as practised in sub-Saharan Africa. For the changing focus and orientation of African social geography in the post-Second World War period reflects close parallels with shifting definitions prominent in the heartlands. Thus in broad terms the three phases of social geography as identified by Eyles (1981) for the Anglo American realm can be identified in the African context; the 'environmentalist' phase is prominently displayed in the largely descriptive and localised 'man-land' studies of the 1940s and 1950s, while the 'morphological/quantitative' phase of the 1960s is to be seen in studies of the social aspects of the 'geography of modernisation' while the 'radical' phase, which in the West was associated with the rediscovery of poverty and the onset of the 1970s recession, is reflected, albeit unevenly, in the study of the socio-spatial aspects of African underdevelopment and its offshoots.

No academic specialism operates in isolation. The phases of development in post-war social geography reflect wider changes in the orientation of the parent discipline and indeed changes within the social sciences generally. The exogenous determinants of African social geography, then, have not emanated exclusively from Western social geography, but also reflect the influences of cognate fields, in particular of development geography which itself reflects the paradigmatic shifts associated with changing approaches to the study of development problems in Africa and other third world regions; shifts which, again until recently, have been closely associated with initiatives from Western scholars (Slater, 1974; Mobogunje, 1980, Ch. 1). A threephased typology of change can be identified for post-war development geography; Porter and de Souza (1974) label these phases, 'man-land', the 'organization of space' and 'historical dialectics'. These labels are immediately suggestive of close parallels with the changes in African social geography. Not least of these parallels is the recognition that the first two phases reflect a subject matter as well as a methodological shift, while the latter embraces the traditional concerns of the respective fields within a radically different methodological and conceptual framework.

In the original 1938 edition of Lord Hailey's An African Survey, geography was summarily dismissed with a brief consideration of the physical structure of the continent. Gilbert and Steel (1945, 22) were incensed by this disregard, particularly, perhaps, because the 1940 Colonial Development and Welfare Act had made half a million pounds available for social science research in the colonies and the allocation of this finance was clearly, in part, to be determined by the priorities identified in Hailey's Survey. However, influenced no doubt by the protests of Gilbert, Steel and others, the 1956 revision of Hailey's work extended a belated recognition, identifying 'the uses of geography' as "... providing a comprehensive survey of the interaction of social, economic and political factors with the physical environment. Geography now has a claim to be regarded also ash an applied science, in view of the aid which it can give to town planning and to the solution of many problems connected with the utilisation of land" (Hailey, 1957, 22),

Taking a cue from Hailey, this chapter focuses on two aspects of sub-Saharan social geography: ecological studies, particularly those that concentrate on population-environment relations and the study of urbanisation and urbanism. While these topics do not exhaust the scope of African social geography they do reflect general trends and have the additional advantage of indicating in perhaps the clearest manner the most likely future developments in the field, for it is in relation to these issues that, in recent times, some of the most innovative work has been completed.

Social Geography and the Ecological Tradition

While the influence of determinist thought in geography had clearly receded by the end of the Second World War, the relationship between 'man' and 'environment' remained a major concern of the discipline, particularly in the geographical study of the tropics. Indeed the 'man-land' theme, in various guises, has retained a predominant in African social geography to the present time. As Porter and de Souza have noted (1974, 73), prior to 1960 geographers studying underdevelopment were preoccupied with the description of landscape and little concerned with social change and development. In social geography this was reflected in the concern with data collection, field survey and mapping (Steel, 1948; Trewartha and Zelinsky, 1954). This predominantly descriptive and empirical approach was encouraged by an adherence to the Hartshornean paradigm of 'areal differentiation' and indeed much of the work in these early post-war years was incorporated into revised editions of regional texts (e.g. Fitzgerald, 1961) as well as providing part of the basis for a new spate of regional geographies (Buchanan and Pugh, 1955; Boateng, 1959; Church, 1964; Kimble, 1960; Sillery, 1961; Hance, 1964; Jarrett, 1962). As late as 1960 pleas were still being made for more descriptive, inventory work (Steel, 1960) and at least one social geography text of the later part of the sixties used this appeal as a justification for publication (Kay, 1967). The most useful contribution of this phase of 'man-land' studies was idenitifed by Thomas as providing a corrective to some of the prevailing myths about the nature of African society.

> Much of the recent efforts of anthropologists, historians, geographers, sociologists and others have been expended in an attempt to correct the misinformation of the human aspects of the former Dark Continent (Thomas, 1965, 250).

The 'winds of change' (Macmillan in Gould, 1961) that blew through Africa in the late 1950s and early 1960s were, in retrospect, less dramatic than was thought at the time. But the attainment of political independence by many African countries was accomplished by high expectations of development and growth. It was the assumption of observers and researchers in Africa that with the aid of Western technology and influence these expectations could be fulfilled. The pessimism of an earlier period, most clearly expressed in the work of Pierre Gourou (1953) who stressed the problems posed by inhospitable tropical environments and the traditionalism of the native people, was replaced by a growing awareness of the capacity of rural Africans to deal with their environments and, from historical studies, an awareness of their adaptability and willingness to change. While social geography remained for the most part wedded to man-environment studies on a local scale, during the 1960s these studies "placed a greater emphasis on an adaption and adjustment to potentialities in the environment and attempted to discover how particular groups of people organised their thoughts about how the environment might be used" (Porter and de Souza, 1974, 73).

These changes, while reflecting a response to external social and political circumstances, were in the case of African social geography mediated through developments in Western social geography and development studies respectively; we are here referring, on the one hand, to the increasing importance of behavioural and perceptual studies (Kirk, 1951; Gould and Ola, 1970) and, on the other, to the adoption of modernisation perspectives (Soja, 1968; Gould, 1970; Riddell, 1970).

It is a curious irony, but an exact reflection of the prevailing structure of communications, that the contribution of Kirk's seminal paper, published initially by the Indian Geographical Society, had to be transmitted with considerable delay via the European heartland before its impact was felt in Africa. The adoption of a perceptual, decision-making focus in African social geography led to numerous studies in the 1960s indicating historically and in a contemporary context how African populations had adapted to and changed their environments. There were close relationships here with other social scientists, particularly anthropologists whose work was developing in a similar direction (Goldschmidt, 1966; Edgerton, 1971; Porter, 1965; Allan, 1965; Newman, 1970; Knight, 1971; 1974).

The influence of development studies was less direct. Quantification and model building which became so much part of the 'geography of modernisation' leading to what Slater has called an 'inverted methodology' (1975), had only a peripheral impact on the study of population-environment relationships in Africa. But the penetration of modernisation ideology, in the sense of an adherence to the concept of dual society and the identification of the Western norm as the model of development, was profound.

The collection of papers brought together by Mansell Prothero, <u>People and Land in Africa:Readings in Social Geography</u> (1972), illustrates the prevailing influences and represents a good cross-section of the kind of work completed during the 1960s. Though retaining a considerable emphasis on the importance of the limitations imposed by the physical environment, and thereby challenging directly the contentions of people such as Ester Boserup

(1965), these studies went a long way to disposing of the myth of traditionalist and conservative peasant and supporting the view of peasants as 'efficient but poor' (see Harriss, 1982, part 3). Yet many of the studies remained at essentially a descriptive level of analysis and at best provided informative case studies. As Prothero recognised in his concluding chapter, studies of 'population-land' relationships lacked an articulated conceptual framework in which the significance of the research could be assessed. Prothero tentatively outlined a model which could be used for this purpose and it is here that the influence of modernisation theory shows through most clearly. Prothero's framework "envisages development along a continuum from traditional to modern, from extensive to intensive, from fluid to fixed circumstances" (Prothero, 1972, 332). What precisely these terms mean is unclear, but the more than passing resemblacne to the terminology employed by Rostow (1960) and indeed Parsons (1960) is sufficient to suggest their philosophical origins. The attachment of the authors in the Prothero collection of modernisation theory with the exception of Barbour (1959), Kay (1965) and Gould (1963), is less explicit but nonetheless clearly signalled.

However, by the end of the 1960s, after 10 years of political independence, having demonstrated the adaptibility and flexibility of African rural dwellers, the population-environment school within social geography was hard put to explain the failure of modernisation and the lack of progress in the African countryside. No longer able to cite traditionalism, conservatism and lack of ambition among African peasants, overpopulation and too rapid rates of population increase were used as explanations. Alternatively, recourse was made to the traditional academic position of pleading ignorance in the absence of enough knowledge accompanied by a call for more research; both were employed by Prothero (1972; 332 and 334).

However, traditionalist explanations were not to be so easily put aside. They emerged in a somewhat modified form in, for example, Gould's contribution to the Prothero collection. In his analysis Gould suggests that Ghanaian peasants have hit upon optimum cropping patterns by chance, by a long process of trial and error; implicitly, at least, the capacity for logical and rational decision-making requiring conscious choice is thereby denied to the Ghanaian farmers. By extending this line of reasoning it is concluded that the changes necessary for modernisation to Western norms would required considerable time, while hard-working but irrational peasants adapt and learn. Such views were presented more forthrightly and more offensively by some writers:

> ...plenty of Africans are less fastidious, taking what they fancy from the foreigner and also keeping what they fancy of their own, much as children do when faced with a supply of new toys which they have not space to stow (Kimble, in McEwan and Sutliffe, 1969, 314).

In the late 1960s and early 1970s the terminology of ecological studies underwent a marked change; 'man-land' and 'population-land' were replaced with a declared concern for 'society-environment' relations. The new terminology reflected the growing influence of political economy approaches and in particular marxist methodology. These influences again emanated from Western, albeit radical sources, and were brought to bear on African social geography predominantly through its links with development studies.

One of the principal attractions of the political economy approach was that it offered the kind of conceptual framework which Prothero, in an entirely different tradition, was attempting to articulate; the kind of framework which, in contrast to modernisation theory, offered some explanation of the disappointments of two decades of development. Thus, in their consideration of these changes, O'Keefe and Wisner (1975, 229) talks of a 'reattachment to reality'. This was a reference, on the one hand, to the shift among some geographers and other social scientists away from the abstract modelling so often associated with modernisation perspectives and, on the other hand, in relation to social ecological studies, the recognition of two important issues; first, that local ecological problems are not satisfactorily explained by reference to local events and processes and that explicit consideration of external factors is required and second, that peasant farmers have a depth of understanding of local environments hitherto unappreciated.

Swift (1975) in his analysis of desertification in the Sahel sums up the first of these issues and in the process demonstrates in practice the kind of interdisciplinary perspective on ecological problems that earlier researchers has merely reflected upon,

> economic stagnation or recession and ecological destruction have been the result of the way in which Sahelian subsistence economies have been increasingly moved towards a market economy, in which they occupy the most peripheral position, a result also of the way in which centralised bureaucratic power has replaced local sources of power and the way in which fragmented modern science has been unable to grasp the ecologically integrated nature of traditional ecosystemms (Swift, 1975, 176).

The second of the above issues (peasants' understanding of their environments) represents an extension of the insights revealed in some of the 'behavioural' studies of the later 1960s (Newman, 1970; Knight, 1974; and especially Richards, 1979 and Karimu and Richards, 1981), and displays a sensitivity towards and an awareness of the importance of local knowledge systems which are not immediately apparent to researchers coming from different cultural and social environments.

In broad terms the significance of the adoption of political economy approaches in the 1970s has been the recognition that 'man-land' relations can only be understood through a prior examination of 'man-man' relations; this is the significance of the use of the 'society-environment' terminology in that 'society' implies social relations.

Social Geography and Urban Studies

Though the extent of urbanisation in pre-colonial Africa has at times been seriously underestimated (Davidson, 1970), it is clear that the colonial period was a time of unprecedented urban development (Hance, 1970). However, for the most part colonial administrations in Africa discouraged African urban settlement on a permanent basis and the level of urban expansion that might have been anticipated given the level of new town creation was curtailed. In the post-independence period the obstacles to African urban settlement were removed and today the rate of urbanisation in Africa (6% per annum) is the highest for any world region; Africa, however, remains the least urbanised of the continents. The relative lateness of African urban expansion explains, in part, the paucity of urban research especially in the 1950s and 1960s when compared with output on urban themes in Latin America and Asia.

The interest of urban social geographers in Africa in the immediate post Second World War year focused predominantly on 'site and situation'. This focus reflected the already noted dominance of the 'man-land' paradigm in geography at that time and despite Steel's 1963 claim that the character of town sites had important social consequences, the general orientation of these locational studies was to illustrate the potential and/or limitations of site characteristics for the physical growth of urban places (e.g. Whittlesey, 1941; Munger, 1951; Jarrett, 1956; Hance and Van Dongen, 1958).

Writing in 1965 Thomas observed

> urban geography in Africa has progressed further on the study of the location of cities than on any other aspect. The site and situation of every town of moderate or large size has probably been investigated by one or more geographers (p. 261).

At the end of the 1950s the interests of African urban social geography shifted towards a concern with the internal morphology of urban places; for example, Larimore (1958) on Kampala, Walmsley (1957) on Nairobi, de Blij (1962) on Lourenco Marques and Lloyd (1962) on Yoruba towns. Studies of this type continued to appear intermittently through the 1960s and into the early 1970s (e.g. Ojo, 1966; Krapf-Arkari, 1969; McMaster, 1968; de Blij, 1968; and Kay, 1967), and were accompanied by other studies which focused on specific spatial characteristics such as population density (Trevellion, 1966) and ethnic/racial divisions (Hallman and Morgan, 1967; Kimani, 1972; Paden, 1971). As with site and situation studies these morphological studies were largely descriptive and only the more recent analyses display a serious concern with the social processes which gave rise to the observed spatial patterns (e.g. Hurst, 1975; Cohen, 1974). Hurst's work in that it employed statistical procedures and adopted an assimilation and, implied, unilinear model of development, was symptomatic of the increasing influence of modernisation perspectives.

Urban places had a crucial role to play in the perspectives of the modernisation school of the late 1960s and early 1970s. The urban systems of Africa were regarded as providing important channels through which the artifacts of civilisation and development could be diffused from core areas to peripheral areas; artifacts which included not only consumer goods and some producer goods, but also appropriate attitudes, values and work disciplines necessary for progress along Western lines. The large cities, in particular, had a special role to play in this process as 'centres of change'; the locus for the initial adoption of modern (Western) attitudes and value systems and for the disregarding of outdated and outmoded traditions.

Soja (1968), after Deutsch (1963), used the term 'social mobilisation' to describe

> ... the general process which underlies modernisation. It promotes the weakening of traditional forms of organisation and behaviour and provides avenues for alternative means of regrouping and restructuring transitional society within a modern framework (p. 3).

Deutsch disaggregates 'social mobilisation' to identify the specific issues involved,

> ... change of residence, of occupation, of social setting, of face to face associates, of institutions, roles and ways of acting, of experiences and expectations, and finally of personal memories, habits and needs, including the need for new patterns of group affiliation and new images of personal identity (quoted in Soja, 1968, 3).

There are clear similarities here with a sociological tradition which at this time (mid and late 1960s) was having a marked effect on the development of Western social geography. Especially important were the works of Oscar Lewis, Louis Wirth and Daniel Moynihan but also relevant was that of Shevky and Bell and to a lesser extent of Robert Park and associates. The specific link was the adoption of a unilinear evolutionary concept, where the 'folk society-urban society', 'rural-urban continuum' of the sociologists paralleled the 'traditional-modern continuum' of the modernisation school; the end product of the evolutionary process was in both cases what Dike (1979) has called 'Euro-American industrial culture'. The perspectives of these two traditions was given explicit expression in the work of development sociologists such as Gutkind, 1970; Mitchell, 1969; Southall and Gutkin, 1957.

In African social geography these influences were most clearly manifested in the studies of rural-urban migration published in the early and mid 1970s (e.g. Pryor, 1975; Harvey and Riddell, 1975; Gould and Prothero, 1975; Ominde and Ejiogu, 1972; Caldwell, 1969; Mabogunje, 1970). Kosinski and Prothero, for example, identify, after Zelinsky (1971), a five stage mobility transition model in which social, economic, demographic and political changes are characterised as moving from a 'pre-modern' to a 'super-advanced' society. In these studies also the quantification and modelling procedures which characterised this stage in the development of Western social geography (Eyles, 1981) and development studies were apparent, as were, to a lesser extent, the influences of behavioural, decision-making approaches (e.g. Garbett, 1975).

Beyond studies of rural-urban migration, urban social geography in Africa in the 1970s focused principally on the problems of the urban environment, especially housing and service provision (Abioden, 1976; Brand, 1976; Kay and Smout, 1977; Hinderick and Sterkenberg, 1975; Peil, 1976; Hake, 1977; Sada, 1977). There were few sub-Sarahan equivalents of Abu-Lughod's (1969) social area analysis of Cairo. In these problem-oriented studies modernisation perspectives were implicitly rather than explicitly demonstrated. This reflects, in part, the continuing dominance of earlier morphological traditions in that the concern was to demonstrate the emerging socio-spatial pattern rather than to investigate process. Indeed when process was given consideration it was narrowly conceived, confined to the analysis of demographic variables; high rates of natural increase and rapid rural-urban migration sufficed to explain problems of housing and service provision. Thus, even when analysis occasionally appeared to contradict modernisation perspectives, as in Van Velsen's (1975) interpretation of squatter settlements not as the apogee of the urban housing problem but as a potential solution, the challenge was directed at the practical detail of housing policy rather than at the philosophy lying behind that policy.

Radical perspectives emanating from political economy have yet to seriously influence urban social geography in Africa. Riddell's (1982) re-evaluation of work on West African migration remains one of the few examples of an explicit attempt to adopt such perspectives (see also, Stren, 1982; Doherty, 1985). A research agenda for urban social geography comparable to that drawn up by O'Keefe and Wisner in 1975 (see also Richards 1980) for social ecology, has yet to be identified. In this respect O'Connor's (1983) recent book on The African City is a profound disappointment in that he rejects the potential application of radical perspectives which have provided the basis for innovatory work elsewhere in the third world (see Santos, 1975, and Slater, 1978; 1984 on Latin America and McGee, 1976, and Drakakis-Smith, 1981 on Asia) preferring instead to stress the specificity of the African context.

While O'Connor's emphasis on the specificity of African urbanism is to a degree to be welcomed, he provides no conceptual or theoretical framework in which such specificity can be evaluated. Mugubane (1979, 46) argues that invariably such an atheoretical approach lends support to the view of urban autonomy, leading to

> a general failure to recognise that the behaviour and attitudes of Africans are not the cause of the kind of urbanisation that Africa has experienced but rather the observable effects of social forces that initiated and shaped the processes of urbanisation itself.

Magubane is here of course referring to the colonial and neo-colonial links which have characterised relationships between Africa and the advanced capitalist countries; the city for Magubane is "an historical configuration reflecting the particular political economy and prevailing class relations of an historical epoch" (p. 32).

While Magubane follows an established political economy argument in calling for an examination of more embracing forces, Dike (1979) in an argument reminiscent of that put forward by Richards (1980) for ecological studies, suggests an approach more sensitised to the role of traditional culture in shaping the nature of African urbanism. Dike argues, in a manner which extends rather than contradicts Magubane, that the hitherto held assumption of the traditional giving way to the modern needs to be seriously revised to allow for for the possibility that "imposed (Western) norms can be indigenised" (p. 20). The relevance of these arguments for urban social geography in Africa has yet to be appreciated and developed.

Conclusion

In 1945 Gilbert and Steel, arguing from the premise that considerations of social welfare were as important as economic growth in the development process, suggested that social geography in Africa, albeit within confines of colonially defined objectives, had a significant and positive contribution to make in identifying problems and formulating policy. As the famine which is presently ravaging Ethiopia threatens to engulf the rest of the Sahel for the second time in 10 years and as the over-crowded and ill-serviced urban slums and squatter settlements continue to expand at an unprecedented rate, it is clear that social geography in Africa has failed to meet Gilbert and Steel's aspirations. There are many possible reasons for this, but part of the explanation has to be found in the nature of social geography as practised in Africa during the post Second World War years.

Hamstrung by inappropriate conceptual frameworks ('man-land' and 'modernisation') derived from alien sources, African social geography has had little opportunity to develop its own distinctive characteristics attuned to the social conditions of African societies. The solution to this problem is not to indigenise African social geography in the sense of requiring the withdrawal of Africanists, nor is it to be found in the kind of false humility espoused by O'Connor (1983, 325) and his associated retreat into ideographic study. Rather, there is clear need for the wholehearted adoption of perspectives which, whilst sensitised to African society and culture, are also capable of embracing the specificity of the African predicament within a wider conceptual framework which permits comparison and evaluation.

Such sensitised perspectives are emerging, but slowly. The previously cited works on 'society-environment' relations by O'Keefe, Wisner, Richards and Karimu remain the best examples. Additionally, recent studies on women's role in the process of social reproduction and their hitherto underestimated role in the direct process of commodity production, augur well for future developments (Momsen, 1984; Spiro, 1984). Significant contributions from African social geography on class formation to match that of African sociologists and political scientists (Cohen, 1981; Sandbrook, 1982; Shivji, 1975) or that produced by geographers working in other areas of the third world (e.g. McGee, 1976) have yet to emerge, except in an historical context (see Crush and Rogerson, 1983).

The basis of much of this work is to be found in 'political economy' which, though having an alien origin, unlike its predecessors, is self-consciously aware of the spatial and temporal conditionality of any explanatory framework and explicitly rejects the universalism associated with both environmentalist and modernisation perspectives. The relevance of an understanding of the social geography of African society is as apparent today as it was in 1945. African social geographers to the extent that they reject the old paradigms and embrace the new, have a unique contribution to make to the unravelling the complex interrelations that make up Africa's development problematic. In this sense the metaphor applied by Skinner (1974) to social anthropology, though it may exaggerate the process, is relevant: African social geography needs to die 'phoenix-like' in order to rise again with a new life form tailored to the present needs of African society.

References

Abiodun, J.O.	1976	Housing problems in Nigerian cities <u>Town Planning Review</u> 47, 339-47
Abu-Lughod, J.	1969	Testing the theory of social area analysis: the ecology of Cairo, Egypt <u>Amer. Socio. Rev.</u> 34, 189-212.
Allan, W	1965	<u>The African Husbandman.</u> Barnes and Noble
Barbour, K.M.	1959	Irrigation in the Sudan: Its growth, distribution and potential extension <u>T.I.B.G.</u> 26. Reprinted in Prothero (1972) pp. 121-139
Boateng, E.A.	1959	<u>A Geography of Ghana</u> C.U.P.
Boserup, E.	1965	<u>The Conditions of Agricultural Growth</u> Aldine
Brand, R.	1972	The spatial organisation of residential

areas in Accra *Economic Geography* 48, 384-96

Brand, R. 1976 The urban housing challenge in Knight, C.G. and Newman, J.L. (eds) *Contemporary Africa* Prentice-Hall

Buchanan, K.M. & J.C. Pugh 1955 *Land and People in Nigeria* University of London Press

Caldwell, J.C. 1969 *African Rural-Urban Migration: the Movement to Ghana's towns* C. Hurst, London

Church, R.J. Harrison et al 1964 *Africa and the Islands* Longmans

Cohen, A. (ed) 1974 *Urban Ethnicity* Tavistock

Cohen, R. 1981 Resistance and hidden forms of consciousness among African workers *Rev. of African Pol. Econ.* 19, 8-22

Crush, J. & C. Rogerson 1983 New wave African historiography and African historical geography *Progress in Human Geography*, 203-231

Davidson, B. 1970 *The Lost Cities of Africa* Little, Brown and Co. (2nd Edn)

de Blij, H. 1962 The functional structure and Central Business District of Lourenco Marques, Mozambique. *Econ. Geography* 38, 56-77

Deutsch, K. 1961 Social mobilization and political development *American Political Science Review* LV, 493-514

Dike, A.A. 1979 Misconceptions of African urbanism: some Euro-American notions in R.A. Obudho and S. El-Shahks (eds) *Development of Urban Systems in Africa* Praeger, 19-27

Doherty, J.	1985	Housing and Development in Freetown, Sierra Leone *Cities* February
Drakakis-Smith, D.	1981	*Urbanisation, Housing and the Development Process* Croom Helm
Edgerton, R.B.	1971	*The Individual in Cultural Adaption* University of California Press
Eyles, J.	1981	Social geography in R.J. Johnston et al. (eds) *A Dictionary of Human Geography* Blackwell
Fitzgerald, E.	1961	*Africa* (9th Edition revised by W.C. Brice) Methuen
Garbett, G.K.	1975	Circulatory migration in Rhodesia: towards a decision model in Parkin (ed) pp 113-125
Gilbert, E.W. & R.W. Steel	1945	Social geography and its place in colonial Studies *The Geographical Journal* CVl, 118-131
Goldschmidt, W.	1966	*Comparative Functionalism: An Essay in Anthropological Theory* Univ. of California Press
Gould, P.	1961	*Africa: Continent of Change* Belmont
Gould, P.	1963	Man against his environment: a game theoretic framework *A.A.A.G.* 53 Reprinted in Prothero (1972) pp. 320-329
Gould, P. & D. Ola	1970	The perception of residential desirability in the Western Region of Nigeria *Environment and Planning* 2, 73-87
Gould, W.T.S. & R.M. Prothero	1975	Space and time in African population mobility in Kosinski and Prothero (eds) pp. 39-49

Gouron, P.	1953	The Tropical World: Its Social and Economic Conditions and its Future Status John Wiley
Gutkind, P.C. (ed)	1920	The Passing of Tribal Man in Africa E.J. Brill, Leiden
Hailey, W.M.	1938	An African Survey Royal Institute of Inernational Affairs, London
Hailey, W.M.	1957	An African Survey: Revised Royal Institute of International Affairs, London
Hake, A.	1977	African Metropolis: Nairobi's self help City Chatto and Windus
Hallman, D.M. & W. Morgan	1967	The City of Nairobi in Morgan, W. (ed) Nairobi: city and region O.U.P.
Hance, W.	1964	The Geography of Modern Africa Columbia University, Press
Hance, W.	1970	Population, Migration and Urbanisation in Africa Colombia Univ. Press
Hance, W. & I. Van Dongen	1958	Dar-es-Salaam: the port and its tributary area A.A.A.G. XLVIII, 419-34
Harriss, J.	1982	Rural Development Hutchinson
Harvey, M.E. & J.B. Riddell	1975	Development, urbanisation and migration: a test of a hypothesis in the third world in Kosinski and Prothero (eds) pp. 51-65
Hinderick, J. & J. Sterkenberg	1975	Anatomy of an African Town: A socio-economic study of Cape Coast, Ghana University of Utrecht Press
Hurst, M.A.	1975	The distribution of migrants in Kampala in Parkin (ed) pp. 319-336

Jackson, P. & S. Smith	1984	*Exploring Social Geography* Allen and Unwin
Jarrett, H.R.	1956	Some aspects of the urban geography of Freetown, Sierra Leone *Geographical Review* XLVI, 334-54
Jarrett, H.R.	1962	*Africa* London House and Maxwell
Jones, E. (ed)	1975	*Readings in Social Geography* O.U.P.
Jones, E. & J. Eyles	1977	*An Introduction to Social Geography* O.U.P.
Karimu, J.A. & P. Richards	1981	The northern area integrated agricultural development project: the social and economic impact of planning for rural change in northern Sierra Leone *Occasional Paper Series* No. 3 Department of Geography, SOAS, University of London
Kay, G.	1965	Resettlement and land use planning in Zambia: the Chipangali Scheme *Scottish Geographical Magazine* 81. Reprinted in Prothero (1972) pp. 284-297
Kay, G.	1067	*A Social Geography of Zambia* University of London Press
Kimani, S.M.	1972	The structure of land ownership in Nairobi *Canadian J. of African Studies* 6, 379-402
Kimble, G.	1960	*Tropical Africa* 3 vols. Twentieth Century Fund
Kimble, G.	1965	Some problems of social change in McEwan, P.J. & R. Sutcliffe (eds) *The study of Africa* Methuen, 305-314
Kirk, W.	1951	Historical geography and the concept of the behavioural environment *Indian Geographical Society* 25, 152-60

Kosinski, L.A. & R.M. Prothero (eds) 1975 People on the move: Studies on Internal Migration Methuen

Knight, C.G. 1971 Ethnography and Change J. of Geography 70, 47-51

Knight, C.G. 1974 Ecology and Change: Rural Modernization in An African Community Academic Press

Krapf-Akrari, E. 1969 Yoruba Towns and Cities Clarendon Press

Larimore, A.E. 1958 The Alien Town: Patterns of Settlement in Busoga, Uganda Res. Paper No. 55. Univ. of Chicago, Department of Geography

Mabogunje, A 1970 Systems approach to a theory of rural-urban migration Geographical Analysis 2, 1-18

Mabogunje, A. 1980 The Development Process Hutchinson

Magubane, B. 1979 The City in Africa: Some Theoretical Issues in R.A. Obudho and S. El-Shahks (eds) pp. 31-53

McGee, T. 1976 The persistence of the proto proletariat: occupational structures and planning the future of third World cities P.I.G. 9, 1-38

McMaster, D.N. 1968 The colonial district town in Uganda in Beckinsale, R.P. and Houston, J.M. (eds) Urbanisation and its Problems Blackwell

Mitchell, J. 1969 Social Networks in Urban Situations Manchester Univ. Press

Momsen, J.H. & J. Townsend 1984 Women's role in changing the face of the developing world Papers of Women and Geography Study Group, I.B.G. Durham

Munger, E. 1955 Geography of Sub-

		Saharan Race Relations in Haines, C.G. (ed) *Africa Today* John Hopkins Univ. Press
Newman, J.L.	1970	*The Ecological Basis for Subsistence Change Among the Sandawe of Tanzania* National Academy of Sciences, Washington D.C.
Obudho, R.A. & S. El-Shahks (eds)	1979	*Development of Urban Systems in Africa* Praeger
O'Connor, A.	1983	*The African City* Hutchinson
Ojo, G.	1966	*Yoruba Culture: a geographical analysis* University of London Press
O'Keefe, P. & B. Wisner (eds)	1975	Land Use and Development *African Environment* Special Report 5
Ominde, S.H. & C.N. Ejiogu	1972	*Population Growth and Economic Development in Africa* Heinemann
Paden, N.J.	1971	Communal competition, conflict and and violence in Kano In Melson, R. & Wolpe, H. (eds) *Nigeria: modernization and politics of communalism* Michigan State University Press
Parkin, D.	1975	*Town and Country in Central and Eastern Africa* International Africa Institute
Parsons, T.	1960	*Structure and Process in Modern Societies* The Free Press, Glencoe
Peil, M.	1976	African Squatter Settlements *Urban Studies* 13, 155-66
Porter, P.	1965	Environmental potentials and economic opportunities: a background for cultural adaptions *American Anthropology* 67, 409-20
Porter, P. &	1974	The Underdevelopment

A. De Souza		and Modernization of the Third World <u>Resource paper</u> No 2 Commission on College Geography, Association of American Geographers
Prothero, M. (ed)	1972	<u>People and land in Africa South of the Sahara</u> O.U.P.
Pryor, R.J.	1975	Migration and the process of modernisation in Kosinski and Prothero (eds) 24-38
Richards, P.	1979	Community environmental knowledge in African rural development <u>I.D.S. Bulletin</u> 10(2) 28-36
Richards, P.	1980	Ecology in African Rural Development - a dissenting view, Paper presented at African Studies Association, Univ. of manchester, September
Riddell, J.B.	1970	<u>The Spatial Dynamics of Modernisation in Sierra Leone: Structure, Diffusion and Response</u> Northwestern Univ. Press
Riddell, J.B.	1980	Beyond the description of spatial pattern: the process of proletarianization as a factor in population migration in West Africa <u>Progress in Human Geography</u> 5, 370-392
Rostow, W.	1960	<u>Stages of Economic Growth</u> Cambridge Univ. Press
Sada, P.O.	1977	Environmental sanitation in urban areas of Nigeria <u>Nigerian Geographical Journal</u> 20, 13-25
Sandbrook, R.	1982	<u>The Politics of Basic Needs: Urban Aspect of Asaulting Poverty in</u>

		Africa Heinemann
Santos, M.	1975	*The Shared Space* Methuen
Shivji, J.	1975	*Class Struggle in Tanzania* Heinemann
Sillery, A.	1972	*Africa: A Social Geography* Duckworth (2nd Edn.)
Skinner, E.P.	1974	*Africa Urban Life: The Transformation of Ouagadougu* Princeton Univ. Press
Slater, D.	1974	Contribution to a critique of development geography *Canadian J. of African Studies* 8, 57-95
Slater, D.	1975	The poverty of modern geographical enquiry *Pacific Viewpoint* 16, 159-76
Slater, D.	1978	Towards a political economy of urbanisation in peripheral capitalist societies *Inter. J. of Urban and Regional Res.* 2, 26-52
Slater, D.	1984	Capitalism and Urbanization at the Periphery - problems of interpretation and analysis with reference to Latin America. Paper presented at I.G.U. Pre-Congress Symposium, Lille, France, August
Soja, E.	1968	*The Geography of Modernisation in Kenya* Syranian Univ. Press
Southall, A.W. & P.C. Gutkind	1957	*Townsmen in the Making* East African Institute of Social Research
Spiro, H.	1984	The Role of Women in Farming and Trading in Oyo State, Nigeria, in Momsen and Townsend (eds), 63-94
Steel, R.	1948	The population of Ashanti: a geographical analysis *Geographi-*

		cal Journal CXII, 64-77
Steel, R.	1960	An Inventory of Land and People J. of African Admin. 12
Steel, R.	1963	African urbanisation: a geographical Viewpoint in Urbanisation in African Social Change Centre of African Studies, University of Edinburgh, 7-13
Stren, R.C.	1982	Underdevelopment, Urban Squatting and the State Bureaucracy: A Case Study of Tanzania Canadian J. of African Studies 16, 67-81
Swift, J.	1975	Desertification and man in the Sahel African Environment Special Report 5, 171-178
Thomas, B.E.	1965	Geography in P. Lystad (ed) The African World Pall Mall Press, 245-270
Trevallion, B.	1966	Metropolitan Kano Newman Neame, London
Trewartha, G. & W. Zelinsky	1954	Population patterns in tropical Africa A.A.A.G. XLIV, 135-62
Van Velsen, J.	1975	Urban squatters: problem or solution in Parkin, D. (ed) Town and Country in Central and Eastern Africa International African Institute
Walmsley, R.W.	1957	Nairobi: the geography of a new city Eagle Press, Nairobi
Whittlesey, D.	1941	Dakar and other Cape Verde Settlements Geographical Review XXXI, 609-38
Zelinsky, W.	1971	The hypothesis of the mobility transition Geographical Review 61, 219-49

Chapter Ten

Research in Social Geography: South Asia

Robert W. Bradnock

'Everyone but geographers knows what geographers do' wrote William Bunge (1983). 'They explore and map.' Supporting his argument that geography is a field subject with characteristically sharp polemic, Bunge was giving voice to a widespread uncertainty over recent directions in geographical research.

That uncertainty is perhaps as relevant to the understanding of research in social geography as it is of other fields in the discipline. The changes that have transformed every aspect of geographical research over the last twenty years have inevitably posed problems for the cohesiveness of research which itself lacks a simple sharp focus. By definition social geography reflects the diversity of the world's social and environmental features. South Asia alone is an enormously rich field for research with a remarkable blending of cultural traditions within its wide variety of physical and economic environments. Yet in many respects social geography as a distinct sub-discipline is in its infancy. Part of the difficulty in assessing progress in social geography in the sub-continent lies in the wide range of relevant research which has been carried out since Independence in 1947, and yet the relatively small amount of that research which has fallen formally into the field of social geography. Indeed, much relevant research has been carried out not only by geographers, but by anthropologists, sociologists, economists and historians, whose own focus of interests may well be elsewhere but whose work sheds important light on questions of concern to the analysis of social geography.

It is impossible to encompass all the relevant work of the last forty years or so. However, in this Chapter I will trace the main features of the subject's development during the period and the chief sources which have fed into the developing perception of the field of social geographical research. Over the last ten years there has been a succession of review articles addressed to the range of progress in geographical research in South Asia. The latest of these is Schwartzberg's discussion of progress in South Asian geography, but they stretch back through the decade to Sopher's challenging call to geographers to 're-discover India', published as a contribution to a wider reassessment of the work of 'geographers abroad' (Sopher, 1973). In between on-going assessments have appeared both from within India and from outside. The reviews of the Indian Council for Social Science Research (1979), under the editorship of Moonis Raza have been complemented by those of Bhatt (1980), Sopher (1980), Murton (1980 a,b), and Noble and Dutt (1982).

The frequency of review articles by geographers working on India may suggest that they are prone to almost yogic contemplation. However, despite the extent of reviewing relatively little direct attention has been paid to social geography. This is not entirely surprising in view of the enormous range of material with which any review of South Asian geography as a whole must come to terms. As Schwartzberg argued:

> materially impoverished though it presently may be, South Asia has, for more than two millenia, stimulated an incredible output of literature relating to its lands, peoples and genres de vie ... While only a small part of that is of more than passing utility for the study of human geography, the resultant corpus is nevertheless quite impressive.

Within this body of research material a substantial proportion is relevant to social geography, including some of the most recent research. None the less there are still important lacunae.

Some of these gaps are geographical in the strictly literal sense. Despite the range of detailed case studies which have been produced by geographers studying India their spread is very uneven. Harriss (1977, 31) showed how studies of agrarian change in India, as reflected in one of the most influential journals, Economic and Political Weekly, showed an extremely patchy distribution in the decade up to 1975. As Figure 10.1 shows, the position up to the end of 1983 was similarly uneven.

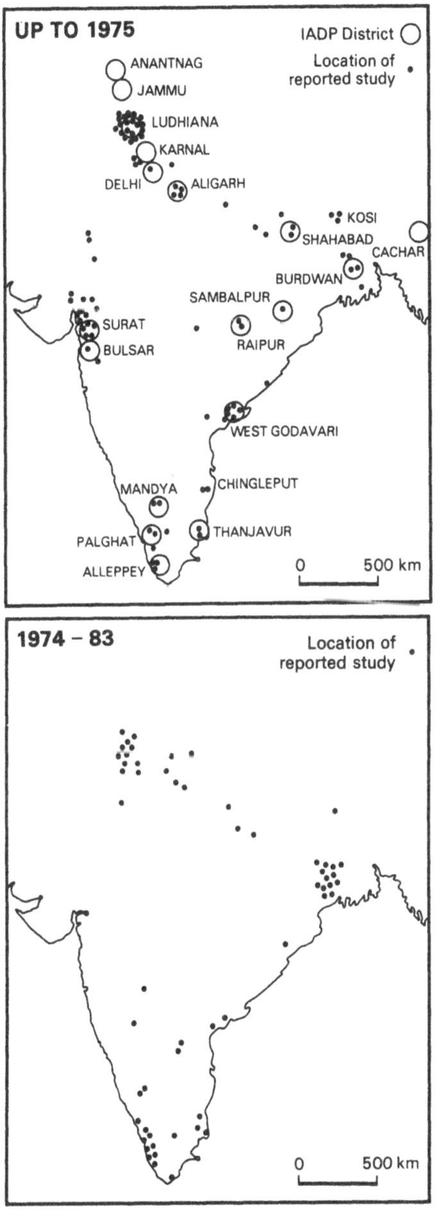

Fig. 10.1 The Spatial Bias of Research in India
Source : Research reported in the 'Economic and Political Weekly'

The two States of West Bengal and Punjab accounted for no less than 22 of the 58 published studies in this latter period. Of the twenty-two studies that related to the four Dravidian States of the south, nine related to the smallest State, Kerala, while only three concerned Karataka, which is five times Kerala's size.

Such measures are of course inherently crude. Many of the papers published in Economic and Political Weekly are of only peripheral relevance to social geography. Nonetheless, the problem of extrapolating argument from a narrow empirical base remains fundamental. Even where a substantial body of research has been carried out, as for example in the very wide range of village studies with which Indian geographical research is liberally supplied, there remain questions as to the general applicability of their findings. As Manor argued (1976) in a review of Epstein's (1973) important contribution to studies of village social and economic change, it is extremely difficult to generalise even at the level of South India on the basis of individual village studies.

In some respects the difficulties of research in social geography in South Asia are increased by the wealth of secondary material and official statistics that are available in comparison with most other regions in the developing world. Schwartzberg has listed the most important sources of data relevant to geographical research in South Asia. India in particular has among the most detailed and sophisticated data sets outside the industrialised world, ranging from extraordinarily detailed village records to macro-level State and national statistics. However, as Sopher argued vigorously in 1973, while the availability of such statistical information is of enormous potential value, it is also open to widespread abuse, nowhere more than in analysing some of the features of social pattern and change that are the substance of social geography in the sub-continent. Thus some of the work on urban social patterns carried out in India during the last fifteen years appears to owe more to the availability of the hard-ware and software to process statistical information than to the rigorous testing of soundly based hypotheses or to a sensitive appraisal of factors not amenable to statistical analysis, however sophisticated.

There is a sense in which Bunge's general strictures on modern geography apply to a degree to research in social geography. Certainly mapping and exploration still have a part to play in interpreting social patterning and organisation, and the regional contrasts in such structures. In South Asia the challenge in this respect still remains enormous. It was Carl Sauer who said "show me a geographer who does not need maps constantly about him, and I shall doubt whether he has made the right choice in life" (1956, 298). Echoing and extending that view Stoddart (1981, 295) has exclaimed "I do not understand that attitude to maps which requires them to be locked up in drawers like mediaeval manuscripts. Maps are to be used, cut up, discarded if necessary; above all, they are to be made".

This message has particular relevance to India, where the quite remarkable map resources of the sub-continent, added to substantially since Independence, remain tragically unappreciated in official circles and underused in many fields to which they have a vital contribution to make. Despite the wide ranging output of Atlas maps published by the National Atlas Organisation of India, the Census of India and others, and despite the highly sensitive and sophisticated Historical Atlas of South Asia (Schwartzberg, 1978), large scale maps remain subject to restrictions on use that continue to inhibit their wider appreciation and therefore their value.

The problems with social geographical research in South Asia do not end with the difficulties - and opportunities - presented by diverse cultural and physical environments, nor with the reliability and availability of data. It is also true that the frame of reference from which geographers have approached the subject matter of social geographic research has varied widely. Still arguably the most influential is that of the regional school, whose work is exemplified by the pinnacle of regional geographies, Spate and Learmonth's India and Pakistan (1968). This was followed by an equally wide ranging regional analysis edited by R.L. Singh (1971).

Given the current weakness of the 'regional' tradition in British geography and the ease with which it was almost totally displaced from British and North American university degree courses some two decades ago the importance of the regional approach to contributions in diverse geographical fields, including that of social geography, may come as something of a shock. While a geographer reading the literature of other disciplines may be uncomfortably aware that the regional dimension is either ignored or deemed insignificant, the integration of the various systematic threads of geography into a coherent regional analysis is a comparative rarity. To Fisher (1977, 90) the purpose of regional research was clear:

> However many regional appraisals geographers may produce, our purpose in doing so is not simply to regard each as an end in itself, like a picture hung on a wall for one's own aesthetic contemplation, but rather to use them as a series of individual case studies, which, treated comparatively, can aid our understanding of the age-old and on-going process of the interaction of man and his regionally diverse territorial environment.

Such an aim would have an obvious bearing both on the subject matter and the approach to geographical research, and it remains the case that even where alternative approaches to that outlined by Fisher have guided research it often remains necessary to disentangle those elements that are distinctively concerned with social geography from a wider whole. The extent to which relevant social geographical research is made less accessible by the range of contexts within which it is discussed is increased by the ever widening breadth of publications relevant to geographical research in the sub-continent. Without making by any means an all-inclusive survey it is possible to name at least a dozen journals in the English language alone which carry research of interest to social geographers. In addition there are those in French, German and Russian, and a growing body of material in Japanese. It is beyond the scope of any individual to keep track of all this material. In addition to the major North American and European sources there is also a rapidly growing body of literature being published in India itself. While this remains very uneven in quality, the recent formation of the National Association of Geographers, India, with its associated journal, the Annals of the Association of Geographers, India, has had an encouraging three years since first publication in 1981.

While we may agree with Bunge that the origins, growth and continued vitality of geography depend on field work, the nature and contribution of that field work varies enormously according to the scale at which it is carried out. In common with anthropology, with which some aspects of social geography have much in common, much research has been carried out at the village level. The number and scope of such village surveys has been discussed in detail by Das Gupta (1977) and Connell (1976). Even making use of advanced statistical techniques Das Gupta found difficulty in making use of many of these studies on a comparative basis as frequently the detailed assumptions underlying them or the methods of data collection were not compatible. Indeed, according to Manor (1976, 156) the renowned social anthropologist M.N. Scrinivas argued at a seminar in Bangalore that "the time had come for village level studies to be de-emphasised in favour of research in wider spatial areas".

At the other extreme are geographical studies which have relied entirely on Census data as the raw material for analysing regionally varied patterns of population and related socio-economic phenomena (Berry & Rees, 1969; Berry & Spodek, 1972; Brush, 1968). Sopher (1973) was particularly critical of such approaches when they appeared to rely on a theoretical base imported from outside India and when they employed little or no testing in the field. Even where sensitively carried out, studies at the broad regional scale inevitably differ radically from micro-level studies, both in terms of the hypotheses that can appropriately be tested and of the featues of social geography that can be analysed.

With this contextual background in mind, it is poossible to review the major fields of research in social geography in South Asia in sharper perspective.

Regional Geographies: Social Pattern in an Environmental Context

The classic post-war regional geography of South Asia, Spate's India and Pakistan (1954) subsequently revised in co-authorship with A.T.A. Learmonth, marked a unique achievement. In its all-embracing frame, it attempted to analyse a comprehensive range of themes some which form a central part of social geography in the sub-continent. In its attempts to examine the expressions and development of South Asia's contrasting cultural life - its religious framework, its caste and social organisation, the physical forms taken by underlying social structures whether in terms of rural settlement types or of urban development - India and Pakistan followed the integrating ideal of that French school of geography which sees the whole as an indissoluble sum of the parts. Encompassing a historical background from the pre-historic to the relatively recent colonial experience, indelibly etched on the face of every part of South Asia, Spate and Learmonth, and Farmer writing on Ceylon (subsequently Sri Lanka) developed a comparative view of south Asia's regional diversity which, building on the earlier work (published in 1982) of Geddes, may never be matched.

However, its very breadth of scope meant that some major themes could be explored at only a relatively superficial level. Some of these - the role of caste and clan in population, settlement and migration in India, for example - were taken up in the second major work of regional geography, R.L. Singh's India: A Regional Geography (1971). Although it attempted to analyse India's geographical diversity from within, the viewpoints of this volume often seemed closer to those of Western geography than of indigenous insights (see Sopher, 1973).

Always far too sophisticated to fall into the traps of environmental determinism or any simplistic analysis, the wealth of historical and geographical evidence brought to bear on phenomena as diverse as house types and village settlement patterns on the one hand and the social factors underlying demographic change on the other sometimes forced these texts to leave more questions unanswered than solved. In so far as they were themselves reviews on a grand scale that was inevitable, but it must also be admitted that the holistic approach itself sometimes seemed to carry with it a superficiality bound up inextricably with the method of analysis. This indeed seems to be partially acknowledge by the fact that Spate and Learmonth's work is divided into a systematic and a regional section, thereby implicitly recognising the difficulty of a complete matching of the two schema. Thus it is to the more explicitly systematic studies in the field of social geography which have begun to emerge in the last two decades that it is appropriate to turn for fresh light.

Cultural Geography

Two major collections of work in progress suggest that cultural geography in India is a particularly active field of research. After the severity of his strictures on the work of Berre et al. (1968, 1972) for their failure to do justic to the cultural dimension in South Asian geographical problems, Sopher (1980a) has set down an alternative set of markers for geographical research. As a collection of diverse essays the book sets out themes which remain to be examined as well as of those in which understanding is already advanced. One might wish that Bhatt's introductory definition of the work of social and cultural geographers had been more roundly fulfilled, for often the physical and environmental dimension scarcely surfaces in the analyses which follow.

That said, it remains an outstanding contribution to the social geography of the sub-continent, and has helped to re-focus attention on the 'Indianness' of some fundamental features of Indian geography. It is a focus that is maintained in Noble and Dutt's set of readings, India: cultural patterns and processes (1982). From the distinctively Indian phenomenon of the sacred cow concept (Simmons 1982) through to the analysis of cultural factors underlying regional contrasts in infant sex ratios and male: female child mortality rates (Miller, 1980) both volumes sustain a high level of analysis, often using innovative methodology, alongside a constant searching out of themes that are relevant in India today (e.g. Hirst, 1982). Some of these themes have a distinctly developmental focus, such as Debysingh's (1980) fascinating analysis of poultry keeping and the factors which inhibit innovation.

Even in research that is defined in systematic raher than in regional terms there is frequently either an acknowledged or an unacknowledged relevance to social geographic themes. Thus as work by Narayanan Nair (1980) shows, it is necessary to look at the question of the implications of the sacred cow phenomenon from an economic as well as from a philosophical or social perspective. Similarly, attempts to understand agrarian change without references to the social geographic framework can at best be partial. Thus Farmer's (1977) project on the green revolution in North Arcot District of Tamil Nadu State, South India, and Sri Lanka, brought together research on both the physical and the sociological sides of the discipline. From the study of the package of changes implied in the term 'green revolution' to the study of individual crops, such as that of Racine (1976) in his analysis of rice cultivation, geographical enquiry has complemented research from other fields.

This complementarity has sometimes been more

apparent than real. Just as one well known interdisciplinary research project was once described as a first class example of cross-sterilisation, so too frequently the lines of communication between disciplines - and indeed sometimes within them - have been desperately weak or non-existent. This is evident particularly in such fields as agrarian reform and marketing geography, where on the one hand political analysis and on the other a bloodless form of economic theorising have too frequently excluded the importance of social geographical factors from consideration.

An example of the positive benefits of interdisciplinary work can be seen in the developing field of medical geography. In view of the continuing relentless rise in population throughout the sub-continent it is remarkable how little geographical research is being undertaken on this most fundamental of geographical relationships, that between man and his essential resource base. This is not to say that little work is being done of relevance to social geography, but it is a field dominated by contributors from other disciplines. Wilde (1982) seems almost alone, for example, in attempting to analyse the determinants of population-resource relationships at the village level, relationships which he showed to be highly dependent on social factors. All too often these factors are ignored by studies which examine population growth as a problem in respect of planning population policy and economic development (e.g. Leontief 1979). There is surely scope for a greater contribution from cultural geographers to the study of demographic change than that which comes simply from demographers alone (see Cain, 1981; Caldwell et al, 1982).

Though geographers have shown relatively little interest in Indian demography, Blaikie (1972) being an important exception, there has been a greater volume of research published in the field of medical geography which does bring together insights from social geography to match those from other disciplines. Medical geography of course has an honourable history, and work has continued in several related fields (see Dutt et al, 1980). With the resurgence of malaria in several parts of South Asia during the 1970s research by Learmonth and Akhtar (1979) has contributed important insights. The spatial incidence of disease has been a natural focus of attention but cultural aspects of disease have also been analysed (Chakravarti, 1982). Chakravarti's paper includes some suggestive maps of food habits, dietary deficiencies and disease.

It has been known for many years that the physical environment may be a factor in disease, through such influences as the mineral content of water supplies. Murali Krishna's (1981) work in Andhra Pradesh is an attempt to examine such factors. Physical factors alone however, are clearly not solely responsible for illness. There are as yet little understood relationships between certain cultural values, health and demographic behaviour. Miller's (1982) research on the neglect of female children carried out in rural north India has established some fascinating patterns and relationships, and Dyson and Moore (1983) have recently attempted to further the analysis of north-south differences in age-sex ratios in terms of kinship systems. Such research represents the blending of complementary disciplines in what can be a highly productive way, and the activity on this research frontier suggests that valuable rewards can be forthcoming (see also Dixon, 1982).

In concluding these comments on the developing contribution of cultural geography it should be recognised that although there is a concern at the appropriatness of explanatory models imported from the western world there is an openness in the work of Sopher and his colleagues both to the historical and the environmental context of cultural geographic research. The geography of religion in South Asia is a field which illustrates this oponness, as does the regional work of Beck (1972; 1979) in the Konku region of Tamil Nadu and Murton's (1977) work on territory, social structure and settlement dynamics in Tamil Nadu before 1800. This work in part reflects the major research scholarship in the field of the historical geography of South India by Burton Stein (1980). While it may remain controversial, Stein's analysis of the regional structure of political power and the ideological concepts through which such power was given legitimacy has opened a new perspective on the historical geography of South India.

The Spatial Organisation of Social Phenomena

There is an inevitable degree of overlap between research summarised above as cultural geography, and the more distinctively spatial research discussed in this section. However, there have been important contributions to the study both of urban and rural settlement patterns which both draws from social geography and contributes to it.

It is true that urban studies in South Asia have received considerable attention from geographers and perhaps a larger amount of comparative work than most other fields of geographical research (see for e.g., Mitra, 1963, 1970; Noble and Dhussa, 1982). However, it is also regrettably the case that such work has frequently been marred by the uncritical borrowing of concepts and tools and from attempts to interpret the results of such analysis without recourse to understanding the historical, social, cultural or economic context.

The value of a sound historical grounding has been exemplified in urban historical research in South Asia for well over a decade. In South India, for example, Lewandowski's (1975, 1980) work on Madras and Madurai has demonstrated the vital need of sound historical analysis for interpreting modern Indian form. Also in South India the work of Frykenberg (1984) on Madras has highlighted key elements of social organisation in the early formation of the spatial structure of those cities. In northern India King's (1975) analysis of Delhi as 'the colonial city' has similarly demonstrated the importance of understanding the political and social history of India's cities if their modern form is also to be adequately analysed, an understanding re-inforced by the work of Harrison (1980).

In contrast to these works geographers have shown an almost blinkered interest in the study of cities as they appear today and of the processes of their frequently rapid transformation. A major area of such geographic research in South Asia and beyond has been in the field of urban morphology. It is perhaps in this field more than any other that the charges of inappropriateness of methodology and interpretation levelled by Sopher (1973) still carry weight. Thus while Brush for example has continued to refine his studies of population density in Indian cities, work which has served as the model for numerous other studies, including for example that of Bangalore by Rao and Tewari (1982), there is still a strong sense that much that is vital to the understanding of cities is lost in the process of reducing their study to the uni-dimensional characteristic of population density.

The possibility of comparison with cities elsewhere in the world gives the attraction of contributing to wider urban theory. However, too often such comparisons are invalidated by exogenous factors. Some of these have been well handled by contributors to Misra's (1978) Million cities of India. Perhaps the most striking of the recent studies of urban morphology are those of Weinstein (1974, 1976) in Madras, who used a factor analytic approach to test some of the conclusions of previous factor analytic studies of Indian cities (Berry and Rees, 1969; Berry and Spodek, 1972). His analysis of the developing morphology of Madras represents a significant enhancement of morphology research, and challenges some of the major conclusions of the earlier research which had suggested a degree of convergence with western urban experience not substantiated by Weinstein's evidence. Other studies have given varying degrees of prominence to social forces in the formation of city structure. Mahadev's (1975) study of Mysore gives a descriptive account of the structure of the princely city, while some of the physical characteristics of Hyderabad's urban landscape have been analysed in terms of underlying social forces.

If urban morphology has long been a focus of Indian urban research so also has the relationship between town and country (see Kapadia and Pillai, 1971). Ellefsen (1960) suggested that around many Indian cities it was possible to trace the evidence of direct urban influence not only on the economic character of the cities' hinterlands but also on their social structures, at least up to a distance of 25 kilometres from the city centres. Bradnock (1974) extended that analysis by means of regression and trend surface analysis, and found it possible to identify clear regional patterns of urban influence on population density and economic specialisation across major regions of South India. Detailed analysis of the links between metropolitan cities and their hinterlands was the hallmark of the Hyderabad Metropolitan Project under the direction of Alam and Khan (1969) and such work has been carried out in many different parts of South Asia.

The relationship between urbanisation and social change has been a further area of research, though one in which sociologists have been more prominent than geographers. Of particular interest, though still relatively under- studied, is the extent to which the rapid growth of cities, which in India alone saw an addition of fifty million people between 1971 and 1981, has stimulated social and economic change. There have been a number of studies on squatter settlements and slums which have strong, though sometimes unstated, geographical implications

(Bannister, 1971; Wiebe, 1975, 1980; Cousins & Goyder, 1979).

While the dynamism of India's urban growth provides an exciting field for research in social geography, rural social patterns have also provided an important range of research areas. The economic and social characteristics of Indian villages have been subject to analysis for decades, Slater (1918) setting an example for field research in South India which has been a model across the sub-continent. Since the late 1950's there has been considerable work on the comparative nature of settlements with particular emphasis on the role of caste as a factor in the organisation of villages. The work of Scarlett Epstein in two villages of Mysore State (1957, 1973) is an excellent example of research which crosses the boundaries of anthropology, economics and social geography. In northern India Singh (1968) and Mukherji (1954, 1982) have contributed to the study of caste and clan movements in the northern plains and the structure of interregional linkages. Alongside these studies has gone research into patterns of migration which again cross the borders between demography, economics and social geography.

Summary

While it is far from complete, the outline of research in social geography in South Asia given in this chapter suggests that despite the fact that scholars from a range of geographical traditions have contributed to work in the field, it retains a strongly idiosyncratic flavour. Some major themes of modern social geography in Europe and North America, such as behavioural geography and welfare geography (see Sivamurthy, 1979), have made only the most limited of entrances onto the South Asian stage. While some of the more recent methodologies, notably quantitative techniques, have made some impression, much of the subject matter remains focussed on problems that are strongly rooted in South Asia's own historical, cultural and physical environment.

References

Alam, M. & Khan, W.	1969	Meropolitan Hyderabad and its region Asia, New Delhi
Bannister, J.	1971	Urban development and

		housing the urban poor - the case of India U.S.AID Memeo
Beck, E.F.B.	1979	Perspectives on a regional culture: essays about the Coimbatore area of South India Vikas, Delhi
Beck, E.F.B.	1972	Peasant society in Konku: a study of right and left sub-castes in South India Vancouver
Berry, B.J.L. & P. Rees	1969	The factorial ecology of Calcutta American Journal of Sociology 74, 445-91
Berry, B.J.L. & H. Spodek	1971	Comparative ecologies of large Indian cities in comparative factorial ecology Economic Geography 47, 266-285
Bhatt, B.L.	1980	India and Indian regions: a critical overview in Sopher, D.E. op. cit. Cornell University
Blaikie, P.	1975	Family planning in India: diffusion and policy Arnold, London
Bradnock, R.W.	1974	The hinterlands of Bangalore and Madras Indian Geographical Journal XLIX, 10-16
Brush, J.E.	1968	Spatial patterns of population in Indian cities, Geographical Review 58
Bunge, W.	1983	Geography is a field subject Area 15, 208-10
Cain, M.	1981	Risk and insurance: perspectives on fertility and agrarian change in India and Bangladesh Pop. & Dev. Review 7, 435-74
Caldwell, J.C. et al	1982	Demographic change in rural South India Population and Development Review 8, 689-728

Chakravarti, A.K.	1982	Diet and disease: some cultural aspects of food use in India in Noble, A.G. & Dutt, A.K. *op. cit.*
Connell, J.	1976	*Migration from rural areas* OUP, Delhi
Cousins, W.J. & C. Goyder	1979	*Changing slum communities* Manohar, New Delhi
Dasgupta, B.	1977	*Village society and labour use* OUP, Delhi
Debysingh, M.	1980	The cultural geography of poultry keeping in India in Sopher, D.E. *op. cit.*
Dixon, R.B.	1982	Mobilizing women for rural employment in South Asia: issues of class, caste and patronage *Economic Dev. and Cult. Change* 30, 373-390
Dutt, A.K., R. Akhtar & H. Dutta	1980	Malaria in India with particular reference to two west central States *Social Science and Medicine* 14D, 317-330
Dyson, T. & M. Moore	1983	On kinship structure, female autonomy and demographic behaviour in India *Population and Dev. Review* 9, 35-60
Ellefsen, R.A.	1960	City hinterland relationships in India in Turner, R. (ed) *India's urban future* Berkeley
Epstein, T.S.	1957	*Economic development in South India* Manchester University Press
Epstein, T.S.	1973	*South India: past, present and future* MacMillan
Farmer, B.H.	1977	*Green Revolution?* MacMillan
Fisher, C.A.	1977	Whither regional geography? *Geography*
Frykenberg, R.E.	1984	The socio-political morphology of Madras: an historical interpre-

		tation in Ballhatchet, K. & Taylor, D. Changing South Asia: city and culture 21-42 School of Oriental and African Studies - Asian Research Service
Geddes, A.	1982	Man and Land in South Asia ICSSR, Concept, New Delhi
Harriss, J.	1977	Bias in perception of agrarian change, in Farmer, B.H. op. cit.
Harrison, J.B.	1980	Allahabad: a sanitary history, in Ballhatchet, K. & Harrison, J.B. The city in South Asia pp 167-96 Curzon Press
Hirt, H.F.	1982	Caste and urban house types in South India: Brahmin houses in Tamil Nadu and Karnataka, in Noble, A.G. & Dutt, A.K. op. cit.
Indian Council Social Science Research	1979	Research in Human Geography New Delhi
Kapadia, K.M. & Pillai, S.D.	1971	Impact of industrialisation on rural social structure Popular Prakasan, New Delhi
King, A.	1975	Colonial urban development Routledge Kegan Paul, London
Krishna, P.M.	1981	Availability of trace elements and incidences of chronic disease: a case study of rural Krishna District Indian Geographical Journal 56, 73-80
Learmonth, A.T.A. & Akhtar, R.	1977	The resurgence of malaria in India Geojournal 15, 69
Lewandowski, S.	1980	Migration and ethnicity in urban India: Kerala migrants in the city of Madras 1870-1970 Manohar, New Delhi
Lewandoski, S.	1975	Urban growth and municipal development in

		the colonial city of Madras 1860-1890 *Journal of Asian Studies* XXXIV, 341-60
Leontief, W.	1979	Population growth and economic development: illustrative projection *Population and Dev. Review* 5, 1-28
Mahadev, P.D.	1975	*People, space and economy of an Indian city an urban morphology of Mysore* Institute of Development Studies, Mysore
Manor, J.	1976	South India: yesterday, today and tomorrow: *Modern Asian Studies* 10, 156-59
Miller, B.D.	1980	The endangered sex: neglect of female children in rural north India, in Sopher, D.E. *op. cit*.
Misra, R.P.	1978	*Million cities of India* Vikas
Mitra, A.	1963	*Calcutta - India's city* New Age, Delhi
Mitra, A.	1970	*Delhi - capital city* Thomson Press, India, New Delhi
Mukerji, A.B.	1982	Cultural ecology of rural settlements in Chandigarh Dun region, in Noble, A.G. & Dutt, A.K. *op. cit*.
Mukerji, A.B.	1954	Jats: a study in human geography *Geographical Review of India* 16, 2, 12-31
Murton, B.J.	1980	The temporal dimensions of famine vulnerability in Ganjam District in the mid-nineteenth century *Indian Geographical Journal* 55, 1-11
Narayan Nair K.	1980	Livestock development: policy and perspectives *E.P.W.* 23, 1005-7
Noble, A.G. & R. Dhussa	1982	Ethnic neighbourhoods in Indian cities: the

		Paras of Dumka and the Bengali Tola of Bhagalpur, in Noble, A.G. & Dutt, A.K. op. cit.
Noble, A.G. & A.K. Dutt	1982	India: cultural patterns and processes Westview Press
Racine, J.	1976	Tradition et modernité en Inde du Sud: deux études rurales en pays Tamoul Centre d'Etudes de Géographie Tropicale
Rao, V.L.S.P. & V.K. Tewari	1982	Urban population densities: models and correlates Geog. Rev. India 44, 2, 1-18
Sauer, C.	1956	The education of a geographer Ann. Assoc. Am. Geographers 46, 297-99
Schwartzberg, J.E.	1978	An historical atlas of South Asia Chicago
Simmons, F.J.	1982	The sacred cow concept: its impact on other bovines, economic life and human genetics, in Noble, A.G. & Dutt, A.K. op. cit.
Singh, K.N.	1968	The territorial basis of town and village settlement in eastern U.P. India Ann. Ass. Am. Geog. 58, 203-20
Singh, R.L.	1971	India: a regional geography Varanasi
Sivamurthy, A.	1979	Spatial pattern of community of property offenders in Madras City Indian Geog. Journ. 54, 79-83
Slater, G.	1918	Some South Indian villages OUP, London
Sopher, D.E.	1973	Towards a rediscovery of India: thoughts on some neglected geography, in Mikesell, M.W. Geographers Abroad: essays on the problems and prospects of research in foreign

areas

Sopher, D.E. (ed) 1980a *An exploration of India: geographical perspectives on society and culture* Cornell University Press

Sopher, D.E. 1980b The geographical patterning of culture in India, in Sopher, D.E. (ed) *op. cit.*

Spate, O & A. Learmonth 1968 *India and Pakistan* Methuen

Stein, B. 1980 *Peasant state and society in mediaeval South India* OUP, Delhi

Stoddart, D. 1981 Geography, education and research *Geog. Journ.* 147, 287-297

Weinstein, J. 1974 *The urban ecology of Madras* Sage

Weinstein, J. 1976 *Modernization and convergence of urban ecological structure* Indian Geographical Society Jubilee Volume Madras

Wiebe, P.D. 1975 *Social life in an Indian slum* Vikas

Wiebe, P.D. 1980 *Tenants and trustees* MacMillan, Delhi

Wilde, C.L. 1982 *Determinants of population-resource relationships at village level* Unpublished Ph.D. Thesis University of London